YALE UNIVERSITY
MRS. HEPSA ELY SILLIMAN
MEMORIAL LECTURES

PROTEIN METABOLISM IN THE PLANT

PROTEIN METABOLISM IN THE PLANT

BY

ALBERT CHARLES CHIBNALL

Professor of Biochemistry
Imperial College of Science and Technology
University of London

NEW HAVEN
YALE UNIVERSITY PRESS
LONDON · HUMPHREY MILFORD · OXFORD UNIVERSITY PRESS
MDCCCCXXXIX

Copyright, 1939, by Yale University Press
Printed in the United States of America

All rights reserved. This book may not be reproduced, in whole or in part, in any form (except by reviewers for the public press), without written permission from the publishers.

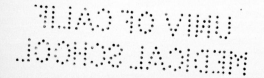

THE SILLIMAN FOUNDATION

In the year 1883 a legacy of eighty thousand dollars was left to the President and Fellows of Yale College in the city of New Haven, to be held in trust, as a gift from her children, in memory of their beloved and honored mother, Mrs. Hepsa Ely Silliman.

On this foundation Yale College was requested and directed to establish an annual course of lectures designed to illustrate the presence and providence, the wisdom and goodness of God, as manifested in the natural and moral world. These were to be designated as the Mrs. Hepsa Ely Silliman Memorial Lectures. It was the belief of the testator that any orderly presentation of the facts of nature or history contributed to the end of this foundation more effectively than any attempt to emphasize the elements of doctrine or of creed; and he therefore provided that lectures on dogmatic or polemical theology should be excluded from the scope of this foundation, and that the subjects should be selected rather from the domains of natural science and history, giving special prominence to astronomy, chemistry, geology, and anatomy.

It was further directed that each annual course should be made the basis of a volume to form part of a series constituting a memorial to Mrs. Silliman. The memorial fund came into the possession of the Corporation of Yale University in the year 1901; and the present work constitutes the twenty-seventh volume published on this foundation.

To the memory of

THOMAS BURR OSBORNE

Gather a single blade of grass, and examine for a minute, quietly, its narrow sword-shaped strip of fluted green. Nothing, as it seems there, of notable goodness or beauty. A very little strength, and a very little tallness, and a few delicate long lines meeting in a point,—not a perfect point, neither, but blunt and unfinished, by no means a creditable or apparently much cared-for example of Nature's workmanship; made, as it seems, only to be trodden on today, and tomorrow to be cast into the oven; and a little pale and hollow stalk, feeble and flaccid, leading down to the dull brown fibres of roots. And yet, think of it well, and judge whether of all the gorgeous flowers that beam in summer air, and of all strong and goodly trees, pleasant to the eyes or good for food,—stately palm and pine, strong ash and oak, scented citron, burdened vine,—there be any by man so deeply loved, by God so highly graced, as that narrow point of feeble green.

JOHN RUSKIN
MODERN PAINTERS, VOL. III, PT. 4, CH. 14.

PREFACE

IN this book my Silliman Lectures are presented in an expanded form. I have, in the first four chapters, discussed the question of protein metabolism in seedlings and have attempted little more than to consolidate the scattered contributions of many of the earlier investigators—particularly Pfeffer, Schulze, and Prianischnikow—and have tried to show that their work cannot be viewed in true historical perspective except in terms of contemporary protein chemistry. As I believe that this section of the book will be of interest to students of both biochemistry and plant physiology I make no apology for the rather detailed nature of the matter presented.

The remaining chapters deal in large part with the proteins of leaves and their metabolism. Here much of the work is admittedly speculative, and in reviewing the present position I have found it necessary in certain cases to address myself to the original workers concerned. I have, accordingly, adopted throughout this section of the book a freer form of writing and have permitted myself a certain latitude not only in the choice of subject matter, but also in voicing my own opinions. At the present stage in the somewhat backward development of this important field of research such liberties may, perhaps, be excused on grounds of expediency, if only in the hope that the book, whatever its merits or faults, may encourage in some measure the wider study and investigation of the chemical mechanisms of plant metabolism. Should this be so I shall feel amply repaid for the time and care entailed in its production.

It gives me pleasure to acknowledge my indebtedness to several friends who have helped in the preparation of this book. Professor H. B. Vickery has supplied timely criticism and unfailing assistance during the writing of the manuscript; in particular, he stimulated my interest in the development of the subject matter of Chapters I and II on historical lines by providing me with long abstracts of certain early papers inaccessible in London. My colleague, Professor F. G. Gregory, has looked over the whole manuscript and has offered criticisms which have been very valuable to me. In thanking both these gentlemen for their very great kindness I need hardly say that they are in no way committed to the views that I have expressed in the text.

Much of the analytical data given in Chapters VI and VII has been provided by workers in my laboratory; Dr. J. W. H. Lugg, in particular, has placed his knowledge of amino acid analysis freely at my disposal and to him, and to other colleagues—Dr. E. J. Miller, Dr. A. Pollard, Mr. M. W. Rees, Dr. G. R. Tristram, Mr. R. G. Westall, and Mr. E. F. Williams—I am deeply grateful. I also offer my appreciative thanks to Dr. J. W. H. Lugg for the very valuable discussion which he has presented in Appendix I, and to Sir John Russell for permission to reproduce portraits of Boussingault and Professor Prianischnikow.

My researches on leaf proteins would have made but little progress in recent years without financial assistance, and I should like to record here my indebtedness to the Imperial Chemical Industries for very generous grants during the years 1929–35 to aid an investigation of forage crop proteins, and to the Agricultural Research Council who have continued since then to support my researches.

Finally, I should like to thank Professor Vickery for his valuable help in the reading and correcting of the proof sheets, and the staff of the Yale University Press for their willing co-operation in the production of the book.

A. C. C.

Imperial College,
  South Kensington; London, S.W. 7.
  16 December 1938.

CONTENTS

List of Illustrations xv

Chapter I. Protein Metabolism in Seedlings.
   Historical Review 1
   1. Introduction. 2. Pfeffer's views on the role of asparagine in protein metabolism. 3. Developments in the theory of protein constitution between 1870 and 1875. 4. Schulze's fundamental chemical analysis of yellow lupin seeds and seedlings. 5. Schulze's objections to Pfeffer's theory of protein regeneration.

Chapter II. Protein Metabolism in Seedlings.
   Historical Review (*Continued*) . . . 26
   1. Borodin's qualified support for Pfeffer's theory. 2. von Gorup-Besanez's postulate concerning the primary products of protein decomposition in seedlings. 3. Schulze's original views on the origin of asparagine in seedlings. 4. Schulze modifies his views on the origin of asparagine in seedlings.

Chapter III. Protein Metabolism in Seedlings.
   Review of the Later Literature . . 54
   1. Introduction. 2. Asparagine as a secondary product of seedling metabolism. 3. Glutamine as a secondary product of seedling metabolism. 4. Arginine as a primary product of protein decomposition in seedlings. 5. The decomposition of the reserve proteins on germination. 6. Protein regeneration in seedlings.

Chapter IV. Asparagine and Glutamine Formation in Seedlings 87
   1. Introduction. 2. The oxidation of amino acids to ammonia. 3. The possible secondary production of the whole asparagine molecule from protein. 4. The accumulation of asparagine and glutamine in plants artificially enriched with ammonia.

Chapter V. THE MECHANISM OF AMINO ACID AND
PROTEIN SYNTHESIS IN PLANTS . . . 104

Chapter VI. THE PREPARATION OF PROTEINS FROM
LEAVES 119
1. Introduction. 2. Review of the early literature. 3. Preliminary investigations on the extraction of proteins from leaves. 4. Definitions of terms used in subsequent discussions. 5. Menke's separation of cytoplasmic and chloroplastic material from spinach leaves. 6. The ether method for preparing cytoplasmic proteins from leaves. 7. The proteins of spinach leaves. 8. Properties and amino acid analyses of the cytoplasmic proteins from leaves of various herbaceous plants.

Chapter VII. THE PROTEINS OF PASTURE PLANTS 145
1. Limitations of the ether method for preparing proteins from leaves; recent researches of Foreman. 2. The ether-water method for preparing mixed protoplasmic proteins from leaves; chemical composition of the chloroplast. 3. Properties and amino acid analysis of the mixed protoplasmic proteins prepared from various pasture plants by the ether-water method. 4. The total protein of the leaf. 5. The nutritive value of the leaf proteins. 6. The cystine content of pasturage and wool production. 7. Relative merits of "true" and "crude" protein applied to forage crops.

Chapter VIII. PROTEIN METABOLISM IN LEAVES . 170
1. Introduction. 2. The non-protein nitrogenous constituents of leaf saps. 3. A general survey of protein metabolism in detached leaves cultured on water. 4. Decomposition of the proteins in detached leaves.

Chapter IX. PROTEIN METABOLISM IN LEAVES
(*Continued*) 189
1. The possible role of asparagine and glutamine in the respiratory activities of plants. 2. The nitrogen-free precursors of asparagine and glutamine in the plant.

Chapter X. THE ROLE OF PROTEINS IN THE RESPIRATION OF DETACHED LEAVES . . . 211

CONTENTS

Chapter XI. THE REGULATION OF PROTEIN METABOLISM IN LEAVES 244

Appendix I. THE IMPURITIES PRESENT IN LEAF-PROTEIN PREPARATIONS AND THEIR BEARING UPON THE ESTIMATION OF SOME AMINO-ACIDS, PARTICULARLY IN RELATION TO THE FORMATION OF HUMIN DURING ACID HYDROLYSIS (BY J. W. H. LUGG) 267

Appendix II. THE ESTIMATION OF NITROGENOUS BASES, DICARBOXYLIC ACIDS AND AMIDE NITROGEN IN THE IMPURE LEAF-PROTEIN PREPARATIONS 280

Appendix III. THE LIPOID FRACTION OF CHLOROPLASTS 284

BIBLIOGRAPHY 286

INDEX 297

ACKNOWLEDGEMENTS

MY thanks are due to the authors and publishers of the following journals for permission to reproduce figures: Revue générale de Botanique, vol. 36, for fig. 1; Planta, vol. 18 (Julius Springer, Berlin), for fig. 2; Zeitschrift für Botanik (Gustav Fischer, Jena), vol. 29, for fig. 3; Proceedings "B" of the Royal Society, London, vols. 117 and 123, for figs. 4 to 8; Connecticut Agricultural Experiment Station, Bulletin 399, for figs. 9 to 19 and the Annals of Botany, new series, vol. 2, for fig. 21.

ILLUSTRATIONS

Figure	Page
1. Daily rates of protein decomposition and asparagine accumulation in lupin seedlings	77
2. Changes in soluble non-protein nitrogen during the forcing of onion bulbs	81
3. Chlorophyll-Protein ratio in leaves of Nasturtium	185
4. Carbohydrate exhaustion and carbohydrate production in detached Barley Leaves	213
5. Carbon dioxide equivalent to the loss of carbohydrate in detached Barley Leaves	214
6. Carbon dioxide production and oxygen uptake in detached Barley Leaves	215
7. Changes in protein and non-protein nitrogen in detached Barley Leaves	217
8. Changes in amino, amide and ammonia fractions in detached Barley Leaves	218
9. Changes in organic solids of detached Tobacco Leaves	222
10. Changes in total sugars in detached Tobacco Leaves	222
11. Changes in starch in detached Tobacco Leaves	222
12. Changes in glucose in detached Tobacco Leaves	222
13. Changes in total nitrogen in detached Tobacco Leaves	225
14. Changes in total protein nitrogen in detached Tobacco Leaves	225
15. Changes in soluble nitrogen in detached Tobacco Leaves	225
16. Changes in amino nitrogen in detached Tobacco Leaves	225
17. Changes in ammonia nitrogen in detached Tobacco Leaves	225
18. Changes in asparagine amide nitrogen in detached Tobacco Leaves	225
19. Changes in glutamine amide nitrogen in detached Tobacco Leaves	225
20. Changes in organic acids in detached Barley Leaves	239
21. Correlation diagram between respiration and protein nitrogen content of Barley Leaves	262

PLATES

Vauquelin	12
Piria	12
Boussingault	12
Pfeffer	12
Ritthausen	28
Ivan Borodin	28
v. Gorup-Besanez	28
Prianischnikow	28
Ernst Schulze	50

PROTEIN METABOLISM IN THE PLANT

CHAPTER I

PROTEIN METABOLISM IN SEEDLINGS HISTORICAL REVIEW

1. *Introduction.*

WHEN seeds are allowed to germinate, one of the products of metabolism which may accumulate in relatively enormous quantities is the amino acid amide asparagine. This substance, discovered in 1806 by Vauquelin and Robiquet (328) who obtained it from the juice of asparagus shoots, crystallises with unusual ease in the very characteristic form of its monohydrate when extracts of plant organs containing a notable amount are concentrated and allowed to stand; it is not surprising therefore that early in the first half of last century asparagine should already have been isolated from a number of plant sources and that chemists should have become interested in its structure, for Vauquelin and Robiquet had shown that it was a neutral organic substance containing nitrogen. As early as 1833, in fact, Pelouze (207) had recognised that it was an amide of the corresponding acid, aspartic acid, but the constitution of the latter remained uncertain until 1862 (Kolbe, 121) and proof that asparagine was its β-amide $HO_2C.CH.(NH_2.)CH_2.CONH_2$ was not finally obtained until Piutti's (218) synthesis of 1887. Meanwhile, however, botanists had realised its importance, for the characteristic crystals of the monohydrate can be readily detected under the microscope in slowly dried sections of asparagine-rich organs, and they had found that under certain conditions it was fairly widely distributed in the plant kingdom. As a soluble organic nitrogenous substance, a possible relationship to protein was soon being discussed, and its metabolism became the subject of more than one extended enquiry.

Of these I prefer to give pride of place to the illuminating work of Piria (216, 217) published in 1844 which was briefly noted by Pfeffer (208) in 1872 and has since then remained forgotten until rediscovered by Vickery *et al.* (336) in 1937. Menici, a pharmacist of Pisa, had had occasion to grow some vetch seedlings in the dark, and from an extract had prepared a small quantity of crystalline material which he submitted to Piria at the University of Pisa for identification. Piria suspected its nature at once, and repeated Menici's experiments on a more extensive scale. The substance was asparagine, and he found that he could isolate equally large amounts whether the seedlings were grown in light or in darkness. Furthermore, as the plants exposed to the light approached the flowering stage, the amount present became very much less and he could detect none at all when the fruit was ripening. Since none could be isolated from the ungerminated seed, Piria concluded that asparagine was produced from a nitrogenous reserve (he suggested a "casein") during germination, that light had no direct influence on its production and that it disappeared again when the plants reached the stage of flowering and fruit formation. These fundamental observations anticipated many of the later researches of Pfeffer, Schulze and Prianischnikow by nearly half a century and the validity of his conclusions has never been contradicted.

Dessaignes and Chautard (65) in 1848 confirmed the presence of asparagine in vetch seedlings grown both in light and in the dark, and in seedlings of nine other members of the *Papilionaceae:* they could detect none, however, in seedlings of buckwheat, pumpkin and oats, even when these were grown in the dark, nor could Pasteur (202) isolate any from 200 litres of sap obtained from vetches approaching the flowering stage. The fact that asparagine which had already accumulated in etiolated seedlings disappeared rapidly when these were exposed to light was first observed by Sullivan (317) in 1858, and was a point emphasised by Boussingault (22) when he summarised his views in 1868.

From Boussingault's work we gain our first insight into the quantitative aspects of chemical changes which take place during germination. Seeds of the pea, wheat, maize and haricot bean were germinated on washed pumice and kept either in light or in the dark for about a month, being watered occasionally with distilled water. Dumas analysis showed that under all conditions the absolute nitrogen content of the seedlings did not change, i.e. no nitrogen was excreted, but combustion analysis of whole seeds or seedlings showed that growth in the dark was accompanied by a large diminution of total carbon, hydrogen and oxygen. The following table illustrates (in the case of the haricot bean) the magnitude of the changes involved, and brings out clearly the differences between the plants grown in light and in the dark. In both instances the relative amounts of hydrogen and oxygen involved are those found in water, and in the darkened plants there was a loss of carbon equal to about 40 per cent of that present initially in the seed.

TABLE 1.

(From Boussingault, 22.)

Haricot bean	28 days' growth	
26.vi.1859	In light	In dark
	$g.$	$g.$
Weight of ungerminated seed	0.922	0.926
" " dried plant	1.293	0.566
Change in organic material	+0.371	−0.360
" " total carbon	+0.1939	−0.1585
" " " hydrogen	+0.0200	−0.0232
" " " oxygen	+0.1575	−0.1781
" " " nitrogen	0	0

Another analysis by Boussingault, this time of maize seedlings grown for a month in the dark, showed that "glucose" and cellulose increased substantially during this period, presumably at the expense of "starch or dextrine," which made the large carbon loss (about 40 per

TABLE 2.

(From Boussingault, 22.)

Maize seed, planted 5.vii.1860 and germinated in the dark for 30 days.

	Total dry weight	"Starch or dextrine"	"Glucose"	Oil	Cellulose	N	Ash	Undetermined
	g.	g.	g.	g.	g.	g.	g.	g.
Seed	0.489	0.362	0	0.026	0.029	0.05	0.009	0.013
Seedling	0.300	0	0.129	0.005	0.090	0.05	0.009	0.017
Change	−0.189	−0.362	+0.129	−0.021	+0.061	0	0	+0.004

cent also in this case) seem even more remarkable. He attributed this loss to carbon dioxide formation during respiratory combustion, and the presence of asparagine in the seedlings showed, in his opinion, that starch, dextrin and oil were not the only materials concerned. Just as part of the protein ingested by an animal (he argued) is modified during respiration to urea, which is excreted, so also in the darkened plant there is a similar respiratory combustion of protein to give, not urea, for the plant cannot excrete, but asparagine, which remains dissolved in the saps and can be again utilised by the plant under the action of light. Asparagine can thus be produced in the roots, stems and leaves, but only in the dark, and it will accumulate only as long as the light necessary for its modification is withheld. In his opinion this explained why he had never found asparagine in the flowering plant, except perhaps in the roots. He had noted, however, that very young plants growing in the light often did contain much asparagine; this he attributed to the proportionally greater development of the roots, which were shielded from light, as opposed to that of the leaflets, which were exposed to it. Boussingault's views are of interest in that they provide us, for the first time, with a *reason* for the transformation of protein to asparagine, and the notion that it might be part of the respiratory mechanism of the plant has often commended itself to later workers. In particular his facile analogy between urea production in the animal and asparagine production in the plant has been vigorously championed in recent years by Prianischnikow. While there is no doubt that it does indeed provide a reasonable interpretation of many observations, I think we should remember that Boussingault produced no evidence whatsoever to show that his asparagine did in fact originate from protein. Pfeffer made this appear fairly certain 10 years later, chiefly through ingenious deductive reasoning, but it was not proved by chemical analysis until 1876 (Schulze and Umlauft, 299).

Side by side with this pioneer work on metabolism chemists were slowly gaining some insight into the nature of the proteins present in seeds. Dumas and Cahours (69) in 1842, using their (then) newly developed method of determining nitrogen, had shown that there were marked differences in their elementary composition and that they were not identical with animal proteins, as Liebig had asserted. In 1855 the botanist Hartig (107, 108) published the results of his elaborate investigations of seeds, in which he showed that a large part of the reserve protein was present in the cells in the form of crystals and grains (aleurone) of more or less definite structure. This was followed from 1860 onwards by the more chemical studies of Ritthausen (243), who devoted himself for many years to the preparation, by extraction with water, alcohol or dilute alkali, of proteins of the highest attainable purity. These were analysed, and the results showed that protein must occur in many diverse forms in the different seeds. His work was criticised in 1876 by Weyl (360, 361) and in consequence fell into partial disrepute, but nevertheless it had a strong influence, as will appear later, on the work of Pfeffer and Schulze.

Hartig (108), who extended his survey to the proteins and other nitrogenous products of seedlings, buds, roots, tubers and the green parts of plants, noticed repeatedly, when he treated sections with absolute alcohol, the crystallisation in characteristic form of a substance (really a group of substances) rich in nitrogen which he called "Gleis." He clearly perceived its significance, for he states, "The apparently general occurrence of this crystallisable substance in all the young cell-tissues, indicates that its solution is the form in which the nitrogenous nutritive substance of plants, formed from the reserve materials, is transported from cell to cell." This important discovery remained unnoticed for some years on account of the peculiar nomenclature employed by Hartig and his rather confused ideas on possible chemical interrelationships. "Gleis," for instance, in one form, such as asparagine, was the "sugar of the gluten meal," but it occurred also in many other forms, certain of which stood in close physiological relationship to the oil!

Beyer (13, 14), however, who made the first, but by no means satisfactory, analysis of the changes which occur during the germination in light of yellow lupins, was aware of Hartig's work, for he found asparagine, which he inferred was "Gleis." He agreed with the latter's contention that it was a suitable substance for the translocation of nitrogen, but queried its origin from the protein reserves. His own experiment showed very little protein breakdown and, since he had found malic acid in the seeds, he suggested that the asparagine might have arisen by condensation of this acid with ammonia. This idea was, in its turn, lost sight of, for the great botanist Pfeffer, who made extensive use of Beyer's analytical data in his investigations published in 1872 (208), was convinced that the asparagine came from protein and ignored the other alternative.

2. *Pfeffer's Views on the Role of Asparagine in Protein Metabolism.*

Pfeffer's monumental researches were concerned with the development of protein in the ripening grain and its metabolism during germination. He used the older, microchemical, technique throughout, but modified the procedure slightly in that the sections of tissue were treated with alcohol to facilitate the crystallisation in their characteristic forms of products such as asparagine dissolved in the saps. He dealt chiefly with members of the *Papilionaceae*, especially the yellow lupin, and disagreed with much of the earlier work of Hartig on the mode of occurrence and distribution of proteins in seeds; in addition he was very caustic about the wide distribution of "Gleis," which he showed must have been in some cases asparagine and in very many others nothing more than potassium nitrate or salts of organic acids such as potassium oxalate. It is not these minor criticisms of Hartig, however, which make Pfeffer's work so valuable but his full and carefully reasoned statement of the position, as he saw it, of asparagine in the protein metabolism of the plant.

Pfeffer considered that the researches of earlier workers had definitely shown that plants cannot assimilate atmospheric nitrogen (Mayer, 165) and that the absolute nitrogen content of the seed does not change during germination unless easily assimilable nitrogen is applied externally (Boussingault, 22). In addition, his own work with many different seeds of the *Papilionaceae,* supported by the quantitative analysis of that of the yellow lupin by Beyer (14), had shown that practically the whole of the nitrogen in the seed was present in the form of protein.

TABLE 3.

(From Beyer, 14.)

Lupinus luteus germinated in light	Per 1000 ungerminated seeds without testas		
	Seeds	Seedlings 1–1½ inches long	Seedlings 2–3 inches long
	g.	g.	g.
Total dry weight	82.1	79.7	77.6
Protein	49.1	46.3	43.1
Asparagine	0.0	0.75	2.6
Glucose and alkaloids	8.45 }	17.09	15.70
Gum	5.54 }		
Cellulose, starch, pectin	8.67	7.71	9.25
Ash	3.38	3.50	3.63

Since, therefore, the small residual non-protein nitrogen was due in large part to alkaloids, and simple forms such as ammonium salts or nitrates had never been detected, it was, in his opinion, quite clear that in this family the asparagine which appears on germination, being richer in nitrogen itself than protein (21.2 per cent as against 15.6–18.2 per cent) could only have been formed at the expense of the reserve proteins of the seed.

Although he did not profess to understand the chemical changes involved, he was emphatic that the asparagine was to be regarded as a direct primary product of the protein decomposition, and quoted in support of this view Ritthausen's recent finding that, on acid hydrolysis, the lupin protein conglutin gave, among other products, aspartic acid, which was the same acid as that obtained from asparagine (247, 249).

To show that such a transformation could be of use to the plant he instanced the conversion of legumin to asparagine without loss of nitrogen.

	125.5 parts of legumin contain*	100 parts of asparagine contain	Difference
C	64.9	36.4	−28.5
H	8.8	6.1	− 2.7
N	21.2	21.2	0
O	30.6	36.4	+ 5.8

This was clearly an oxidation, and the carbon and hydrogen set free could either be used for the synthesis of new nitrogen-free plant material such as carbohydrate, or be burnt to carbon dioxide and water. He deduced from Beyer's analyses of lupin seeds and seedlings that during germination the latter alternative was quite possible, but stated that, in his opinion, respiration might not be quite such a simple matter as this.

The asparagine thus formed in the cotyledons diffused out into the axial organs where, if carbon and hydrogen could be made available from some nitrogen-free source

* From Ritthausen's analysis (246).

such as carbohydrate, or even some nitrogen-poor source, it would be regenerated to protein. This new product, however, would be albumin, which Pfeffer had found widely distributed in the growing parts of plants, and not legumin, which occurs only in seeds. He stressed the importance of carbohydrate in protein regeneration and pointed out that light *per se* could play no such fundamental role as Boussingault had suggested, but merely acted as a promoter of photosynthesis, since plants grown in the light, but in the complete absence of carbon dioxide, remained loaded with asparagine until death (209). In the absence of available carbon and hydrogen, asparagine must accumulate, and this fact, according to Pfeffer, explained the slight divergence in the findings of earlier workers; Piria's experiments with vetch seedlings, for instance, gave as much asparagine in seedlings grown in light as in those grown in the dark because the former had not developed sufficient foliage to provide the necessary supplies of carbohydrate for protein regeneration; Pasteur's flowering plants had already converted all their seedling asparagine to protein, while the asparagine disappearance noted by Sullivan was due to rapid elaboration of carbohydrate through photosynthesis. He disagreed strongly with Boussingault's analogy between urea and asparagine, which he thought misleading because the latter was never excreted and when conditions warranted was always utilised again by the plant, as Boussingault himself and many others had already shown.

Pfeffer considered that this intimate connection between protein, asparagine and carbohydrate held only for seedlings of the *Papilionaceae*. In other families, asparagine seemed to be of less importance; Dessaignes and Chautard, for instance, could find none in seedlings of buckwheat and oats, yet Boussingault had found it in maize and Lermer in barley malt, both of which plants, as does the oat, belong to the *Gramineae*. He himself could find none in seedlings of pumpkin, castor oil, *Specularia speculum* and rape, and only for a very short period in those of nasturtium, *Silybum Marianum* and sunflower, which produced the same amount whether grown in light or in the dark. He concluded therefore that asparagine was not of such widespread occurrence as Hartig had maintained, and suggested that in some plants its place might be taken by a derivative of asparagine, or a totally different type of substance such as solanine, which Boussingault had thought might replace asparagine in the potato. It is interesting to note that a few years later Schulze found the postulated asparagine derivative or rather homologue (glutamine) in all the seedlings mentioned above.

To revert to Pfeffer's own work with the *Papilionaceae;* he had found very little asparagine in the cotyledons unless these became green, i.e. were epigeal as in the lupin, but an abundance in the parenchyma of the cortex and pith,* from which he deduced that the breakdown of protein to asparagine took place in the cotyledons, whence the asparagine was transported in the parenchyma of cortex and pith to the axial organs, and there combined with carbohydrate and other nitrogen-poor material to form protein, or was stored as potential material for this purpose. Strongly influenced as he was in those days by the contemporary work of Thomas Graham, it was perhaps only natural that he should have regarded asparagine, a substance which could diffuse so readily through the cell wall, and which his own observations had convinced him was very mobile, as the ideal medium whereby protein material could be translocated from one part of the plant to another, especially in the *Papilionaceae,* and that he should have assigned to it a role in protein metabolism comparable with that of glucose in carbohydrate metabolism.

The only other simple nitrogenous substances known in 1872 as widespread plant constituents were ammonia and

* Pfeffer appears to have been puzzled that such large amounts of asparagine should remain dissolved in the sap, for he never found crystals in the untreated plant tissue. Schulze and Umlauft discussed this question in 1876 (299). See p. 17.

nitric acid, and he discussed the possibility that these might function in the above-mentioned manner instead of, or together with, asparagine. He admitted that such a role for ammonia was not to be regarded as impossible, since Hosaeus (112) claimed that small amounts were formed from protein during germination, but expressed the opinion that neither ammonia nor nitric acid could replace asparagine since no one had found either substance in seeds, and therefore it was unlikely that they would be formed during the breakdown of the seed proteins. Returning to the subject four years later, in 1876, Pfeffer (210) noted von Gorup-Besanez's work of 1874 (91) showing that protein decomposition in vetch seedlings gave leucine and suggested that this substance, with perhaps other decomposition products not yet identified, might replace asparagine, especially in families other than the *Papilionaceae* in which this particular substance appeared to be of less importance. Bearing all these facts in mind, I do not think it can be maintained that Pfeffer was so biased in favour of asparagine as an intermediary in plant protein metabolism as certain later critics have suggested.

That proteins themselves might migrate in the plant through the thin-walled elongated cells of the vascular bundles, as Sachs (258) had originally suggested, was a possibility that Pfeffer also envisaged; he held in fact that this would probably be necessary in those plants which apparently could not make use of asparagine, and might even take place also in others alongside the translocation of asparagine in the parenchyma of the cortex and pith. Such transport, however, would necessarily be slow, but he thought that diffusion alone might be adequate if the proteins were present as potassium or potassium phosphate salts, since Hoppe-Seyler (111) had shown that potassium albuminate could pass through membranes more readily than albumin itself. In 1876, he amplified this and suggested that the protein might diffuse much more readily as peptone, the enzymes required to produce this having been recently discovered in vetch seeds (von Gorup-Besanez, 93). Finally, he stated that, in his opinion, the question of translocation of protein material in the plant was of importance only in the early stages—the vegetating plant probably supplying its needs from the products of assimilation and salts of ammonia or nitric acid drawn from the soil.

Pfeffer's views received immediate recognition and inspired many investigations planned during the succeeding 20 years. Paradoxically he had, without performing a single quantitative analysis himself, placed the whole relationship between asparagine and protein on a semiquantitative basis, and the rather unsatisfactory position thus revealed attracted the attention of a young chemist, Ernst Schulze, who had that year (1872) been transferred to Zürich as professor of agricultural chemistry.

To appreciate the reason why Schulze found himself forced almost at once to challenge some of Pfeffer's ideas, it is necessary to realise that, during the period 1870–1875, the importance of amino acids in protein structure was for the first time appreciated, and the revolution in thought thus initiated, which culminated about 30 years later in the peptide hypothesis of Hofmeister and Fischer, was one that affected Schulze strongly and immediately, but which came too late to influence the ideas Pfeffer formulated between 1870 and 1872.

3. *Developments in the Theory of Protein Constitution between 1870 and 1875.*

As I mentioned earlier, Ritthausen began the first serious study of the vegetable proteins in 1860. At that time only four amino acids were recognized as hydrolysis products of proteins: glycine had been prepared from gelatine, serine from silk, and both leucine and tyrosine from many different sources. Of these, only the last two were regarded as important. In 1866 (244), he hydrolysed the protein we now know as wheat gliadin and found glutamic acid; two years later (245), on hydrolysis of conglutin, the chief reserve protein of lupin seeds, he ob-

VAUQUELIN

PIRIA

BOUSSINGAULT

PFEFFER

tained, besides leucine, tyrosine and glutamic acid, a new acid (aspartic) which in 1869 he recognised (247) to be the same as that given by the hydrolysis of asparagine. These were important discoveries, but at the time very few investigators had appreciated the significance of amino acids as integral parts of the protein molecule, and it is probable that Ritthausen alone considered the determination of amino acids to be a significant means of characterising proteins. Intelligible hypotheses of protein structure were, of course, not advanced until the end of the century. It was against such a vague background that Pfeffer, between 1870 and 1872, had to formulate his views on the transformation of protein to asparagine and, as I stated before, in justice to him the point needs emphasis, for his paper displays a truly remarkable appreciation of the importance of chemistry in the elucidation of plant physiological problems; indeed I regard it as one of the great pioneering works of plant biochemistry.

Unfortunately perhaps for him, the year 1872 was also a landmark in protein history as it witnessed the publication of Ritthausen's famous book, *Die Eiweisskörper* (248), the source from which Schulze undoubtedly drew his early ideas on protein constitution. Here Ritthausen reviewed *inter alia* the results of his previous 12 years' work, and, when discussing the isolation of aspartic and glutamic acids after hydrolysis, which he admitted could not be quantitative, put forward for the first time the suggestion that the amount isolated was probably characteristic of the protein concerned. He quoted the following figures from his own and Kreusler's work.

Protein	Glutamic acid %	Aspartic acid %
Mucedin (modern wheat gliadin)	25	not determined
Maisfibrin (modern maize glutelin)	10	1.4
Gluten-casein (modern wheat glutelin)	5.3	0.33
Conglutin	3.5	2.00
Legumin	1.5	3.50

He also called attention to the fact that the aspartic acid thus isolated was optically active, as also was that derived from asparagine, whereas the substance made by heating the ammonium salts of malic, fumaric and maleic acids (Pasteur, 203) was inactive; it appeared therefore that the protein molecule must contain either asparagine itself or groups closely related to it. Ritthausen was voicing here for the first time two very important postulates: (1) that the amino acids produced on hydrolysis were derived from structures already formed in the protein molecule, a question which was to be hotly disputed by chemists during the ensuing 20 years, and (2) that asparagine was probably present in the protein molecule.

Indirect support for the latter view was furnished the same year by Nasse (182, 183, 184), who worked out an exact method for estimating the ammonia given on hydrolysis of proteins. The amount formed during alkaline hydrolysis invariably exceeded that resulting from acid hydrolysis, but the rate at which it was given off was rapid at first and then slow; he inferred therefore that the protein molecule contained two ammonia-yielding groups, one of which, containing "firmly bound" ammonia was decomposed only by hot alkali, and then at a slow rate, while the other was decomposed by hot acids and at a rapid rate by hot alkali. He thought that this "loosely bound" ammonia might, among other possibilities, be present in the protein molecule in the form of acid amide groups, and instanced the like behaviour of asparagine.

This suggestion was further developed in 1873 by Hlasiwetz and Habermann (110), who, deducing from their experiments that casein on acid hydrolysis yielded exclusively leucine, tyrosine, aspartic acid, glutamic acid and ammonia, suggested that, in the intact protein molecule, the latter was combined partly with the aspartic acid as asparagine and partly with the glutamic acid as the analogous, but then unknown, amide which they referred to as glutamine.*

* Glutamine was discovered in the juice of beetroot by Schulze and

These conclusions of Ritthausen and of Hlasiwetz and Habermann gave chemists for the first time reasonable grounds for believing that amino acids, with their essential structure unchanged, entered largely into the composition of the protein molecule, and, since glycine and serine were not discovered among the hydrolysis products of vegetable proteins until after the development of Fischer's ester method of analysis in 1901, their composition was considered to be qualitatively similar to that postulated for casein. Schulze certainly accepted this view, and was thereby eventually forced into open conflict with Pfeffer, but he realised, perhaps more clearly than some of his early contemporaries, that leucine, tyrosine, asparagine and glutamine must represent at the most only a part of the intact protein molecule and, in the absence of what he regarded as definite experimental evidence, he was always very cautious when he discussed the possible composition of the remainder. There was, at the time, no suggestion that proteins might be composed exclusively of other (unknown) amino acids; in fact nitrogenous bases such as xanthine and hypoxanthine were regarded as possible constituents from the work of Salomon (264), who isolated them from lupin seedlings. Schulze, when discussing Pfeffer's views at this period, never denied that sugar, perhaps in altered form, might somehow be incorporated in the protein molecule.

4. *Schulze's Fundamental Chemical Analysis of Yellow Lupin Seeds and Seedlings.*

Pfeffer's theory of protein metabolism in the plant (as his contemporaries referred to it) did not meet with the unqualified approval of Schulze in 1872 as it was based in

Bosshard in 1883 (292) and much indirect evidence accumulated later (cf. Osborne, 189) in support of Ritthausen's and of Hlasiwetz and Habermann's suggestions, but they were not experimentally confirmed until quite recent years by the isolation of asparagine from an enzymic digest of edestin (59) and of glutamine from an enzymic digest of gliadin (60).

part on the chemical analysis of yellow lupin seeds and seedlings made by Beyer (table 3), whose methods he considered were in most instances quite unreliable. The seeds had been germinated in the light and, even at the second stage (2–3 inches long), the cotyledons were still enclosed in the seed coat and had thus not yet emerged above ground. The amount of protein breakdown and of asparagine formed was accordingly fairly small and, as the latter was estimated by crystallisation, the yield, in the presence of so much other soluble material, must have been, so Schulze considered from his own experience, very low. Also Ritthausen had shown (248) that the lupin protein conglutin contained 18.4 per cent of nitrogen, so that Beyer's factor for determining total protein (N $\times$ 6.25) was much too high; moreover his fraction "glucose and alkaloids" was particularly misleading in that it could have contained only 0.6 per cent of alkaloid ("bitterstoff" of Siewert [309]) and, being nothing more than an alcohol extract of the seed or seedling, must have contained, among other products, citric acid (Eichhorn, 74), while his "gum" was merely undetermined water-soluble material. Schulze, who was an extraordinarily skilled and shrewd analyst, felt that Pfeffer's conclusions were of such far-reaching importance that they should rest on firmer chemical evidence, and the first task that he set himself at Zürich was to repeat Beyer's work under more stringent conditions.

His results were published in collaboration with Umlauft (299) in 1876. This paper was the first of a long series which appeared during the next 35 years in which he described his researches on many aspects of protein metabolism in the plant, all of which were alike remarkable for the fine technical skill displayed and the close reasoning with which the results were discussed. Many of them were outstanding contributions to protein chemistry, and obtained immediate recognition from those interested, while others, particularly on plant metabolism, appear to have been less widely appreciated, possibly because he was so meticulous in discussing any new suggestion from every angle that there was, from the reader's point of view, a wearisome amount of repetition. A perusal, however, of his papers shows that his caution, nay his timidity, in advancing new views on metabolism was in keeping with his whole attitude towards his own and other peoples' researches, which was that of a chemist who wholly mistrusted speculation that was not based on *ad hoc* chemical evidence, and yet who realised, from his wide experience, how difficult such evidence often was to obtain. For this reason I think the advances he made were, at the time, reflected more in the progress of protein chemistry than of plant physiology, but, by the end of his life, he had brought his labours in the latter field to fruition and had placed the subject of plant protein metabolism (among others) on such a firm chemical basis that his main conclusions are never likely to be disputed.

Schulze and Umlauft (299) germinated yellow lupins in the dark on wire gauze placed above distilled water so that the rootlets were immersed. Samples of the seedlings were collected for analysis after 7 and 14 days, when the hypocotyls were 2.0–2.5 cm. and 7–9 cm. long, and the rootlets were 4–5 cm. and 6–8 cm. long, respectively. Tabulated results in terms of 100 g. of ungerminated seeds without testas are given in table 4. It will be observed that at each stage a considerable loss of total dry weight is recorded, showing that respiration had been very active (no material was lost to the water) and that protein decomposition had been rapid and extensive. This is reflected in an enormous accumulation of asparagine, far greater than the results of previous workers had indicated, and at the second stage it amounted to 22.3 per cent* of the total dry weight of the seedling! As will be

---

* Schulze and Umlauft record that, in 15-day-old etiolated seedlings (the limit of time before death ensued), one quarter of the dry weight of the seedling was present as asparagine. All their asparagine values were obtained indirectly by Sachsse's method (260), but the amount obtained by direct crystallisation was only about 10 per cent less. The

TABLE 4.

(Compiled from Schulze and Umlauft, 299, and Schulze, 268.)

Lupinus luteus germinated in the dark — Per 100 g. of ungerminated seeds without testas

	1 Ungerminated seeds	2 7-day-old seedlings	3 12-day-old seedlings	4 Difference 3 − 1
	g.	g.	g.	g.
Protein	45.07	24.93	11.66	−33.41
Asparagine	0	9.78	18.22	+18.22
Amino acids ("amido-N")* ...	3.06 ⎫		8.82	+ 5.76
Other soluble nitrogenous products*	4.74 ⎭	15.73	11.6	+ 6.86
Fat	7.75	3.95	2.08	− 5.67
Dextrinous material	10.02	0	0	−10.02
Glucose	0	4.51	2.10	+ 2.10
Citric and malic acids	1.92 ⎫	11.35	0.57	− 1.35
Other soluble N-free material ..	3.86 ⎭		3.55	− 0.31
Crude fibre	3.24	4.13	6.47	+ 3.23
Other insoluble N-free material	16.44	9.09	12.67	− 3.77
Ash	2.97	3.48	3.40	+ 0.43
Total weight	100.0	87.4	81.7	−18.3

seen from table 4, 33.41 g. of protein had disappeared, to be replaced (in part) by 18.22 g. of asparagine.

If the observed changes are interpreted in terms of nitrogen instead of weights it will be found that only 63.5 per cent of the lost protein nitrogen has reappeared as asparagine nitrogen. Schulze and Umlauft therefore real-

concentration of asparagine in the 12-day-old seedling sap was only 1.5 per cent and they call attention to the fact (cf. Pfeffer, p. 10) that the value is below that which Plisson (219) found for the limiting solubility of this substance in cold water. In 1880, however, Schulze (270) found 2 per cent! (Cf. table 14, p. 40.) We know now, of course, that the solubility of amino acids is a function of the dielectric constant of the solvent, which is 80 for water, but much higher when amino acids are dissolved in it (Cohn, 55).

* In calculating the weights of these respective fractions I have adopted the factor N $\times$ 6.

ised that the remaining 36.5 per cent must have been transformed into other water-soluble products and, since von Gorup-Besanez (91) had, two years previously, found leucine in vetch seedlings, they suspected the presence of "amides."* They accordingly determined "amido-N" by the method of Sachsse-Kormann† (262). Their results are given in tables 5 and 6, and it will be seen that about one-fifth of the lost protein nitrogen is present in this form. Again following von Gorup-Besanez (93), they searched for peptones by dialysis, and found evidence for the presence of reasonable amounts at stage 2, but less at stage 3 (table 4). They also found small amounts of free ammonia (about 0.25 units in terms of nitrogen), but were uncertain whether to regard this as being a decomposition product of protein or an artifact due to the breakdown of asparagine during the operations incidental to analysis.

Besides these changes in the nitrogenous constituents

* In those days no clear distinction was made between amino acids and amino acid-amides; both were referred to indiscriminately as "amides" and their nitrogen as "amido-N." In Schulze's earlier papers, however, "amido-N" was used in the same sense as the modern "amino-N," although he used the term "amides" for amino acids. Other workers sometimes referred to non-protein nitrogenous products in a loose way as "amides," so that care is necessary in interpreting the results of early papers. I think I am right in stating that the modern term "amino-N," with its present-day significance, did not come into general use until Van Slyke had published his well-known method of estimation.

† Based on the interaction between amino groups and nitrous acid ($KNO_2 + HCl$) to give gaseous nitrogen. The apparatus used was very cumbersome, and the full significance of the results obtained with plant extracts was not realised at the time. The method is, of course, the forerunner of that of Van Slyke. Schulze in 1877 (267) pointed out that unreliable values were obtained with amides unless these were first of all hydrolysed with hydrochloric acid. The abnormalities he observed were explained later, in 1883 (292), when he and Bosshard isolated glutamine and found that it gave nearly double the theoretical value for "amido-N." In skilled hands such as his, the method was undoubtedly capable of giving good results, although the non-removal of ammonia from the plant extract and the use of hydrochloric instead of acetic acid would make the determined values a little too high.

TABLE 5.
(From Schulze, 268.)

100 g. of ungerminated seeds (without testas) gave, after 12 days' germination in the dark, 81.7 g. of seedlings.

Lupinus luteus	Protein-N g.	Asparagine-N g.	"Amido-N" g.	Ammonia-N g.
81.7 g. of 12-day-old seedlings	2.08	3.86	1.47	0.26
100 g. of ungerminated seeds	8.16	0	0.51	0
Difference	−6.08	+3.86	+0.96	+0.26

TABLE 6.
(From Schulze, 268.)

Lupinus luteus	Protein decomposed		In decomposition products				
	Total weight	Total N	Asparagine-N	"Amido-N"	Ammonia-N	Undetermined N	% of total N accounted for
	g.	g.	g.	g.	g.	g.	
10-day-old seedlings	36.13	5.78	3.23	1.17	0.32	1.06	81.7
12-day-old seedlings	38.0	6.08	3.86	0.96	0.26	1.00	83.7

there was, during germination, a rapid utilisation of the nitrogen-free reserves present in the seed. Both fat and organic acids showed a steady decrease, while the dextrinous material had disappeared completely in under 7 days. Glucose, absent from the seed, was present in the hypocotyls at stage 2, but had partially been used up again at stage 3, due possibly, they thought, to cellulose formation. By comparison with the behaviour of other plants, Schulze and Umlauft concluded that, in lupin seedlings, there was a rapid breakdown of protein in the dark because the seed reserves were poor in fats and available carbohydrates, and, in agreement with Pfeffer's theory, it could be held that the large accumulation of asparagine was due to the absence of sufficient nitrogen-free material to regenerate it to protein. Whether carbohydrate or other nitrogen-free material had actually been produced during the protein decomposition—as the theory demanded—and had later been used up in respiration or in the synthesis of products such as cellulose, they could not say. Their experimental findings showed clearly that the lost protein had been transformed partly into asparagine and partly into other products, including "amides" (amino acids). Might not these "amides" contain much more carbon and hydrogen than asparagine?

5. *Schulze's Objections to Pfeffer's Theory of Protein Regeneration.*

Schulze was voicing here for the first time what he was beginning to feel was the underlying weakness of Pfeffer's theory, namely, his tacit assumption that, in seedlings of the *Papilionaceae,* protein breakdown gave rise directly to sugar and large amounts of asparagine, and that protein regeneration was concerned solely with the recombination of these two substances. Two years later (268), he was in a position to formulate his objections in greater detail and to put forward alternative suggestions of his own. Those on the chemistry of protein breakdown —based on the fundamental work of von Gorup-Besanez —were not at first very satisfactory, and the question was to receive his close attention for the ensuing 20 years; it will be convenient therefore if we discuss first of all his objections to Pfeffer's views on protein regeneration.

If, Schulze argued, asparagine accumulates in seedlings because sugar or other nitrogen-free or nitrogen-poor material is not available for protein regeneration, how can one explain its presence in Beyer's seedlings (table 3)? These were germinated in light and the total material lost through respiration was small. Ignoring therefore Beyer's analyses for carbohydrates, which, for reasons stated above, he mistrusted, it seemed to him quite clear that these seedlings could not be suffering from any carbohydrate deficiency. Why then was not the asparagine immediately regenerated to protein?

Again, he had himself germinated lupins both in darkness and in the light, with results that were not in accordance with the theory. Table 7 gives the results of one such experiment. At the end of 5 days in the dark, the testas had been shed and, after 7 days in the light, the cotyledons were green and the first leaflets had unfolded. The axial organs of the seedlings exposed to the light con-

TABLE 7.

(From Schulze, 268.)

Seedlings of *Lupinus luteus*	Cotyledons		Axial organs	
	Total dry weight % of fresh weight	Asparagine % of total dry weight	Total dry weight % of fresh weight	Asparagine % of total dry weight
Grown 5 days in dark, followed by 7 days in the light	13.26	9.75	6.86	27.22
Grown 12 days in the dark ...	11.94	7.93	3.86	29.29

tained nearly twice the total solids (evidence of active photosynthesis) and twice the amount of asparagine of those grown continuously in the dark!

Even more striking were the results obtained with lupins grown in washed sand, at first for 10 days in the dark so that the seedlings became loaded with asparagine (stage 1), and then for 3 weeks (stage 2), and 6 weeks (stage 3), respectively, in the light. Exposure to light had brought about rapid photosynthesis, which is reflected in an increase in total dry weight (per 100 g. of dry seedlings at stage 1) of 65.5 g. at stage 2 and 161.8 g. at stage 3. At stage 2, there was also an increase of 5.4 g. in protein, yet, in spite of this, there was actually an *in-*

crease in asparagine of 3.99 g.! After another 3 weeks, the asparagine content had fallen somewhat, but even so the absolute amount present remained relatively very high.

TABLE 8.

(From Schulze, 268.)

Seedlings of *Lupinus luteus*   Per 100 g. of dry seedlings at stage 1

	Stage 1	Stage 2	Difference 2 − 1	Stage 3	Difference 3 − 2
	g.	g.	g.	g.	g.
Total dry weight	100	165.5	+65.5	261.8	+96.3
Protein	16.94	22.34	+ 5.40		
Asparagine	17.34	21.33	+ 3.99	12.59	− 8.74
Total N	10.76	10.76		10.76	

Schulze concluded from these observations that, in both the early and late phases of lupin-seedling metabolism, the presence of much available carbohydrate did not cause the regeneration of protein from asparagine. The increase of protein shown at stage 2 had not taken place at the expense of asparagine at all—on the contrary, as the data in table 9 show, *both the protein and the asparagine itself* had been synthesised from the other soluble nitrogenous products present and these were known to contain amino acids ("amides") (cf. tables 5, 6). He and Umlauft (299) had already suggested that breakdown of the reserve protein in lupin seeds on germination gave, besides asparagine, amino acids rich in carbon instead of the sugar that Pfeffer's theory postulated; it now appeared probable that these amino acids might be preferentially utilised in the regeneration of protein in the axial organs.

Such a hypothesis was, after all, more in keeping with the complex structure of the protein molecule than that of Pfeffer, for proteins were known to contain a series of nitrogen-containing groups as well as sulphur, and it was difficult to imagine how they could be built up from sugar

TABLE 9.
(From Schulze, 268.)

Seedlings of *Lupinus luteus*	In percentages of total N		
	Stage 1	Stage 2	Difference
Protein-N	25.2	33.2	+ 8.0
Asparagine-N	34.2	42.0	+ 7.8
Other soluble-N	40.6	24.8	−15.8

and one single nitrogenous substance such as asparagine. If asparagine was indeed utilised for this purpose, then it was far more probable, as Mercadante* (163) and follow-

* The object of Mercadante's research was to disprove the suggestion that asparagine formed during the early stages of leguminous seed germination was later transformed, if the young plants were exposed to light, into albumin. He worked with the yellow lupin and the kidney bean and claimed to have shown in each case that, as the asparagine disappeared, no albumin was formed at all but only aspartic acid, succinic acid and ammonia. He considered that these were used with the plastic substances (Pfeffer's term for labile metabolic products) of the protoplasm to nourish the plant. He gave a short account of his methods of isolation but no analyses of his products, and his results were too incomplete to warrant serious discussion, especially as he provided no evidence that the small amount of succinic acid found had actually been formed from the asparagine. Further, as Schulze remarked, his aspartic acid was undoubtedly a decomposition product of asparagine formed through mishandling the plant material during the course of analysis. In tabulated form his results with yellow lupins were as follows:

Seedlings of *Lupinus luteus*	Length of seedling *cm.*	In percentages of total dry weight		
		Asparagine	Aspartic acid	Succinic acid
Grown in dark	18	15.25	trace	
" " "	15	12.30	1.34	much
" " light	9	14.42	trace	
" " "	?	3.23	2.14	1.26

The results with kidney beans were similar. In view of subsequent developments in asparagine metabolism, one must give Mercadante credit for a pioneer suggestion, which his results superficially confirmed. Cossa's paper was very brief and was meant to confirm Mercadante's; it contained very little beyond the statement that vetches one metre high no

ing him Cossa (57) had already suggested, that it was first broken down to malic or succinic acid and ammonia, and that the latter was used subsequently for protein synthesis.

longer contained asparagine, which had been transformed to calcium succinate and malate.

CHAPTER II

PROTEIN METABOLISM IN SEEDLINGS HISTORICAL REVIEW (*Continued*)

1. *Borodin's Qualified Support for Pfeffer's Theory.*

FROM the account given in the previous chapter it would appear that, in 1878, Schulze was of the opinion that protein regeneration in lupin seedlings took place in the first instance from amino acids, but that, at a later stage, and probably more slowly, this could be brought about, in an indirect way, by condensation of asparagine and carbohydrate or other nitrogen-poor material. Furthermore, the accumulation of asparagine was not conditioned by the amount of carbohydrate available. Schulze's deductions were undoubtedly based on good chemical grounds, but from a physiological point of view they were unsatisfactory in that they were restricted to one particular plant—the lupin—and to one period only of its growth—germination. Pfeffer's theory had been deduced from experiments covering a wider field, but this also had been deliberately limited by him to those seedlings in which he could find much asparagine. The time was therefore clearly propitious for an extension or modification of these views, to embrace, if possible, the whole plant kingdom, and the position was fortunately clarified later in the same year by the publication of a commendably brief and brilliant, though in parts highly speculative, paper by Ivan Borodin (21), who was later appointed professor of botany at the Agricultural Institute in St. Petersburg (Leningrad), and who had for some years been interested in the chemical mechanisms connected with respiration.

Borodin pointed out first of all that, although proteins

were generally regarded as the most important compounds of protoplasm, their chemical activities were not yet known. Was it likely, he asked, that they were entirely passive amid the active changes taking place therein? He felt that this could not be true and had, in fact, already suggested (20) that they were the direct agents in respiration. Respiration was, as he had shown, a function of the amount of carbohydrates present, but these substances could play only an indirect role in the regeneration of material required for protoplasm, otherwise it would be impossible to explain Garreau's observation (86) that the young parts of plants, which are richest in protein, have the most intensive respiration, and not those parts richest in non-nitrogenous substances. In animals, the proteins were considered to be the substrates of respiration and he held that the same was true in plants. (This view had, of course, been propounded many years previously by Boussingault, of whose work Borodin seems to have been unaware.) The question therefore arises, what are the decomposition products of protein which, with the carbohydrates, are used again for protein regeneration?

Pfeffer had considered that asparagine was used for this purpose, but only in certain plants, particularly the *Papilionaceae* in which he could find it, and then only during the germination period. Borodin agreed with the postulate but doubted the limitation, for he was convinced that asparagine must be generally present in *all* plants and that the decisive part it played in protein regeneration was universal. Such was the original view of Hartig (108), to which Pfeffer had objected because he could find no asparagine in the buds of certain plants.

To clear up the controversy, Borodin investigated the development of buds in certain woody plants, which process he regarded as physiologically analogous to the germination of a seed. The first experiments were made on branches cut in winter and kept indoors with the cut end in water. Contrary to Pfeffer's experience, asparagine* was found in all parts of the shoots of *Tilia parvifolia*, whether the buds had developed in the light or in the dark. The same was found to be true of shoots from many other plants, in agreement with Hartig.

Pfeffer's negative results therefore required explanation, and the key was found in the following experiments. Borodin was surprised to find that he could obtain no asparagine at all from large twigs of *Lonicera tatarica* which had been cultivated in the dark for some weeks and had produced 5–6 cm. etiolated shoots with flower buds, although *Sambucus* shoots, which were next investigated, gave as usual much asparagine. Two weeks later, however, the *Lonicera* flowering shoots had withered and dropped off, and the small upper axillary buds had developed new etiolated, but purely vegetative, leafy shoots 4–5 cm. long, which contained large amounts of asparagine. He pointed out that these conflicting results could not be ascribed to the different morphological nature of the material, since both leaf and flower shoots of *Syringa vulgaris* yielded asparagine in both cases.

Borodin argued that if, as Pfeffer assumed, asparagine is formed out of protein and can, with carbohydrates, regenerate to protein, asparagine will accumulate only when the second process is the slower (by lack of carbohydrate or slower supply). If, therefore, *Lonicera* has a large carbohydrate reserve, rapidly translocated to the developing buds, asparagine formed will be immediately used up and will not appear until a later stage, when the

* All his experiments were made by the simple Hartig-Pfeffer method of examining sections under the microscope; the sections were dabbed with absolute alcohol and, after two hours, when the alcohol had all evaporated, asparagine was indicated by the formation of its crystals either on the section or at the edge of the cover slide, which was put on after the first moistening (not before, as with Pfeffer). If the asparagine crystals were very minute and were mixed with much foreign material, their change of appearance at 100° (when they lose water) and their insolubility in a saturated solution of asparagine (at the same temperature) were noted.

RITTHAUSEN

IVAN BORODIN

v. GORUP-BESANEZ

PRIANISCHNIKOW

carbohydrates are largely reduced in amount. This explanation was supported by analogous experiments with *Populus tremula,* which produced a rosette of etiolated leaves in the dark; one of these leaves, picked at once, had no asparagine, while another, picked two weeks later, contained considerable amounts. Now carbohydrate supply in cut twigs is limited, but, in the natural state on the tree, carbohydrate will be continually available so that asparagine should not appear. Experiments made in the spring entirely confirmed this view and thus settled the above-mentioned controversy, since Hartig had used cut twigs and Pfeffer normal buds.

Borodin found that normal buds could be divided into three groups: (i) those in which no asparagine was present at all, e.g. *Alnus glutinosa;* (ii) those with a very small asparagine content, e.g. *Quercus pedunculata* and (iii) those with considerable amounts, e.g. *Spirea sorbifolia;* the results varying somewhat from bud to bud even of the same plant and at different stages of development. In all cases, the undeveloped buds contained no asparagine and their development is thus analogous to the germination of seeds.

Those buds showing no asparagine (group i) could be made to produce it, however, if carbohydrate supply be limited, i.e. if the buds were grown on cut branches, or the small flower shoots were detached from the branch. *Alnus* was the most difficult case, and asparagine did not appear until the eleventh day.

The foregoing results showed that, in agreement with Hartig and contrary to Pfeffer, asparagine was of widespread occurrence in developing buds, and, this time in agreement with Pfeffer, that asparagine accumulation therein depended on carbohydrate supply. Furthermore, the asparagine was produced at the place at which protein was decomposed, whereas the carbohydrate required for protein regeneration must have been obtained from the stems. The analogy between the developing bud and the germinating seed was thus very close, and in both cases asparagine could be looked upon merely as a transition substance between the reserve protein of bud or seed and the cells of the meristematic parts. Borodin was certain, however, that the role of asparagine was not so restricted as this, for he could find it under suitable conditions in all parts of the plant at all stages of development. He demonstrated this in the following way. Normal living plants such as *Vicia sepium* seemed to be free from asparagine, for he could detect none in organs such as stems, leaves or unripe pods,* but if a young part of the plant was detached and kept in a moist, dark room for a week, asparagine at once accumulated and tyrosine could be detected in stems, leaf stalk and leaves. The same was true of detached single leaves and unripe pods. Fourteen other plants gave similar results, but a few exceptional cases were encountered in which detached shoots gave no asparagine at all, e.g. various *Cruciferae* and *Labiatae*. These were considered to be merely extremely refractory examples, as the ease with which asparagine appeared varied from plant to plant.

He discussed next the origin of the asparagine. Aware of von Gorup-Besanez's results (to be presented in the next section), he agreed that protein decomposition in the plant gave asparagine, tyrosine (he had repeatedly detected this) and products which he felt were worthy of further investigation; all were, moreover, in his opinion, produced by unknown ferments present and active during the whole life of the plant. Two possible theories accounted for this: (i) the decomposition processes were confined solely to non-nitrogenous substances as long as these were available, the protein not being affected until carbohydrates were lacking and (ii) the protein was always being decomposed, yielding asparagine etc., but could only be regenerated if sufficient carbohydrates are present. The ultimate result would be, of course, the same

* Quantitative analysis by Schulze and later workers has since shown that this is not strictly true.

by either hypothesis, and although data were lacking with which to make a decision Borodin stated that he preferred the second as being simpler and more likely. If it appears remarkable, he added, to assume asparagine formation when none could actually be detected, it may be remembered that it is assumed that respiration continues in green organs under the action of light, although no carbon dioxide appears, and that present even disappears.

Finally, Borodin commented on Schulze's criticism of Pfeffer's view that accumulation of asparagine was due to lack of carbohydrate, since he (Schulze) had found that 10-day old etiolated lupins, when brought into the light, showed in 5 weeks a small increase in both asparagine and protein content concomitant with an increase by nearly one-half in total dry weight. It is wrong, he stated, to assume that only carbohydrates are required (e.g. potassium is necessary) for protein formation, or that every kind of nitrogen-free substance is available for this purpose; starch and fatty oils, for instance, cannot be utilised and it is possible that the only useful form is glucose. Thus potato tubers contain asparagine because the sugar translocated from the leaves is transformed into insoluble starch. Germination, however (cf. Schulze's interpretation of Beyer's results), is primarily a question of the conversion of insoluble reserves into soluble useful material, and asparagine will appear as long as decomposition is quicker than regeneration.

Borodin concluded therefore that Pfeffer's theory could be brought into line with his own and Schulze's findings if it were supposed that :—

(i) Very few of the nitrogen-free substances present in plants, perhaps only glucose, can bring about regeneration of protein from asparagine.
(ii) In certain cases, even glucose may not do this because it is required for other purposes, e.g. respiration, cell growth, etc.
(iii) Breakdown of protein to give asparagine occurs not

only in the cotyledons of legumes but also in all the living parts of all plants.

(iv) In living plant organs, it is probable that protoplasmic activities such as respiration require a continuous breakdown and regeneration of protein. This protein breakdown gives asparagine and, if sufficient carbohydrate (glucose) is available, the asparagine is at once regenerated to protein; if it is lacking, however, the asparagine accumulates.

Pfeffer's reaction to these suppositions was unfavourable, and later, in 1880 (211) he would do no more than admit that some cells might decompose protein and others regenerate it; Schulze, however, agreed that they helped to explain many of his observations and, briefly in 1879 (269) and with a wealth of detail in 1880 (270), he produced much supporting evidence. He pointed out first of all that on general chemical grounds glutamine, an acid amide not yet isolated from plants, but for whose presence therein he already had much indirect evidence, could in questions appertaining to protein decomposition be regarded as equivalent to asparagine, so that Borodin's failure to find asparagine in e.g. various *Cruciferae, Ricinus*, etc., was probably due to the fact that such plants elaborate glutamine instead (he was to prove this within a few years).

Next, he agreed that Borodin's second supposition, if true, might well explain the behaviour of his lupins grown at first in the dark and then exposed to the light. In further support, he produced evidence from experiments with 15-day-old etiolated seedlings of lupin, soya bean and squash showing that enrichment of amides in young plants is, as a rule, the stronger the less nitrogen-free reserve material is present in relation to the quantity of protein. This comes out clearly from the data given in tables 10 and 11.

His experiments with potato tubers, however, were at first sight not so convincing. As was well known these

TABLE 10.

(From Schulze, 270.)

Per 100 g. of ungerminated seeds without testas

Seedlings of	Protein (N × 6.25) g.	Fat (ether extract) g.	Dextrinous carbohydrate g.	Ratio Protein : N-free reserves
Lupin	51.0	7.75	10.02	1 : 0.35
Soya bean	42.69	17.06	11.91	1 : 0.7
Squash	34.25	53.6	–	1 : 1.56

TABLE 11.

(From Schulze, 270.)

Seedlings grown in the dark for 15 days		Per 100 g. of ungerminated seeds without testas Asparagine + glutamine formed (Sachsse) g.	Protein decomposed g.	Protein decomposed in terms of total protein in ungerminated seeds without testas %
Lupin, at temperature of	18–20°	19.43	42.0	82.4
Soya bean "	" 18–20°	9.57	17.7	41.4
Squash "	" 25–26°	5.78	14.8	43.2

contain much starch, which is formed from carbohydrate (glucose?) translocated from the upper parts of the plant. He and Barbieri (289) had found that the tuber sap contained much asparagine and "amido-N" (Sachsse-Kormann), and had isolated small amounts of leucine and tyrosine. Such a nitrogenous mixture was typical of that which he had found in lupin seedlings and he was inclined to agree with Borodin that it probably arose from the decomposition of protein. One may well ask, then, why the glucose flowing in from the upper parts of the plant did not react with the asparagine to regenerate protein? Borodin had explained this by supposing that the glucose was at once condensed to starch, but Schulze pointed out that this explanation assumes that the glucose concentration in the tubers was very low, and he had actually found 0.64 per cent! In spite of this he felt, nevertheless, that Borodin's explanation was probably the correct one, because he could not be certain that his analytical methods would definitely distinguish between glucose and other reducing substances. He was, however, careful to reiterate his former view that protein might also be regenerated from asparagine and nitrogen-poor material such as tyrosine without the intervention of glucose at all.

Henceforth he regarded what he called the Pfeffer-Borodin hypothesis as a simplified but convenient statement of protein metabolism in the plant, and much of his later work was directed towards an explanation of the chemical mechanisms involved, which he foresaw as early as 1880 must be far more complex than either Pfeffer or Borodin had imagined.

2. *von Gorup-Besanez's Postulate concerning the Primary Products of Protein Decomposition in Seedlings.*

In his original paper of 1872, Pfeffer (208) stated that asparagine was to be regarded as the sole primary nitrogenous decomposition product of the reserve proteins of seeds. The first suggestion that this was only partly true came from von Gorup-Besanez, whose fundamental postulate—that, on germination, the reserve proteins of the seed undergo enzymic digestion with the production, as on acid hydrolysis, of several amino acids or acid amides —has been lost sight of in the extensive developments to which it was subjected in later years by Schulze, who, however, was always careful to point out von Gorup-Besanez's priority.

Eugen Freiherr von Gorup-Besanez, professor of chemistry at Erlangen from 1855 until his death in 1878 and a famous exponent, in his day, of zoöchemical analysis (he discovered valine in the aqueous extract of pancreas in

1856), devoted the last few years of his life to an examination of vetch seeds and seedlings. In 1874 (91), he found that the etiolated seedlings, when 2–3 weeks old, contained quite reasonably large amounts of leucine* as well as much asparagine, and remarked that this fact could not fail to be of considerable chemical and physiological interest. A few months later (92) he followed this with the statement that, since he could find no leucine at all in the ungerminated seed, its presence in the seedlings showed that it must have been formed during the process of germination, and that from the seed itself he had extracted a ferment which energetically converted albumin to peptone (93). Finally, in 1877 (94), he remarked that, as he could find asparagine and leucine in the seedlings, he considered it reasonable to suspect the presence of glutamic acid and tyrosine as well. The isolation of glutamic acid had long eluded him, but he had now found it by the method Schulze and Barbieri (287) had applied a few months earlier to extracts from pumpkin seedlings; he had also been able to detect tyrosine by the colour reaction of L. Meyer. Having thus isolated or detected all four of Hlasiwetz and Habermann's protein amino acids (cf. p. 14) he concluded that, during the germination of vetch seedlings, the reserve proteins underwent enzymic digestion to give products which were chemically the same as those given by acid or alkaline hydrolysis.† At the time it was enunciated this postulate was undoubtedly justified.

von Gorup-Besanez died the following year, and further development came from Schulze (268), who had at

* The product usually described in those days as leucine crystallised in nodules and was shown later to be a mixture of leucine and phenylalanine.

† A close translation of his statement reads as follows: "Evidently during the germination of the vetch a splitting of the albuminous compounds of the seed reserves takes place, chemically coinciding with that which occurs during the metabolism of such substances in the animal body and likewise, externally to the living organism, by the action of known chemical agents."

once recognised the importance of the new postulate since he had already obtained much evidence in support of it. He had shown, for instance, that breakdown of protein in lupin seedlings gave no inconsiderable amount of "amido-N" (cf. table 5), and he had actually isolated leucine and tyrosine as well as asparagine; then again, he and Barbieri (287) had found, in addition to these three substances, glutamine (indirectly as glutamic acid) in pumpkin seedlings. But, like von Gorup-Besanez, he could find none of them in the respective seeds.

Furthermore he was, within the next year (1879), to discover phenylalanine, which he and Barbieri (288) isolated, first from lupin seedlings, and then (290) from the hydrolysis products of the lupin seed protein conglutin. Valine was found in lupin seedlings in 1883 (291) and, in the same year, he and Bosshard (292) isolated from sugar beets the long sought and eagerly awaited glutamine. In succeeding years, he was to add most of the other protein amino acids to this list, thereby providing ample proof for von Gorup-Besanez's contention.

With regard to the agents which brought this about, he agreed that these were probably the seed ferments, for he had himself found evidence for peptone in lupin seedlings. He pointed out, however, that other processes could not be excluded, since proteolytic ferments gave peptones and not crystalline products,* and there was still uncertainty as to the real mechanism of asparagine production. It was, in fact, the enormous elaboration of this substance in etiolated lupin seedlings which made him realise that von Gorup-Besanez's postulate, true though it undoubtedly was in principle, could not, without modification or extension, explain all his observations connected with the breakdown of the seed reserve proteins during germination.

* He was apparently unaware of, or did not grasp, the full significance of Kühne's (128) observation in 1876 that digestion products resistant to the further action of pepsin were attacked by trypsin with the production of tyrosine and "leucine" (leucine + phenylalanine).

3. *Schulze's Original Views on the Origin of Asparagine in Seedlings.*

If, he argued in 1878 (268), the seed reserve proteins on germination are decomposed as von Gorup-Besanez suggests to give amino acids and acid amides, then the relative proportions in which these are formed should be the same as on acid or alkaline hydrolysis and would be determined solely by the constitution of the particular proteins concerned, with the difference that the amides will, in the two latter cases, be decomposed to their respective acids and ammonia. Yet this postulate was contrary to his experience: the conglutin of lupin seeds, for instance, like the proteins of pumpkin seed, yielded much leucine on acid hydrolysis, but this material was found only in traces in their respective seedlings. Then again, conglutin yielded very much more glutamic than aspartic acid on acid hydrolysis; more glutamine than asparagine might therefore be expected in lupin seedlings. This again was not true, for in no case had he been able to recognise glutamine at all. In pumpkin seedlings on the other hand, he could detect much glutamine and very little asparagine, although the seed protein gave about equal amounts of the respective acids. How were these differences to be explained?

Pfeffer had shown that the products of protein decomposition in seedlings could be again utilised for regeneration. In Schulze's opinion, however, a mixture of such products was not necessarily required; any one of them might suffice,* since Knop and Wolf (120) had grown water-cultured rye plants successfully when nitrogen was supplied entirely in one form, e.g. tyrosine or leucine, and Bente (12) had grown maize similarly on asparagine. Asparagine and glutamine were no doubt extensively utilised by many plants for this purpose and his own work had shown that under certain conditions (cf. table 9) lu-

* For Schulze's views on the synthesis of protein from one single organic nitrogenous substance, i.e. asparagine, see p. 52.

pin seedlings used amino acids such as tyrosine and leucine preferentially. Protein decomposition products which are found in any quantity therefore are those either unsuitable for regeneration or are utilised only very slowly for this purpose; in lupin seedlings, asparagine clearly comes into this category, just as glutamine does in pumpkin seedlings. It was not a question of simply rebuilding with the decomposition products from the reserve proteins of the seed—the new protein in the growing parts was albumin and some different process must be operating.

Although such a hypothesis would explain many observed facts, it would not wholly account for the enormous accumulations of asparagine that are sometimes encountered in lupin seedlings, since Ritthausen had obtained only 2 per cent of aspartic acid from conglutin. If allowance be made for inadequate methods of isolation, a higher value must be taken and an exaggerated one would be, say, 16 per cent, in which case not more than 20 per cent of the nitrogen from decomposed conglutin should be found as asparagine and 80 per cent in the form of other products. If the latter be used preferentially for protein regeneration, this relationship would alter in favour of the asparagine, but to attain that found in 15-day etiolated seedlings (60 per cent) very active regeneration would be necessary and this was contrary to his observations, for, in such seedlings, the total protein was only about one-fifth that present originally in the seed, and some conglutin remained undecomposed. Protein regeneration had therefore not been very extensive and only a part, probably a small part, of the asparagine could have come directly from conglutin. The remainder must have originated in some other way, and the data presented in table 12 suggested that it was at the expense of the other decomposition products.

Both Hartig and Pfeffer had found much more asparagine in the axial organs than in the cotyledons; he had confirmed this in 1876, and had now investigated the point

TABLE 12.

(From Schulze, 268.)

Seedlings of *Lupinus luteus*	N of decomposed protein converted into asparagine-N
	%
Grown 4 days in dark	45.7
Grown 7 days in dark	56.7
Grown 12 days in dark	63.5
Grown 10 days in dark and then 14 days in dull light*	73.4

in some detail. His new data, presented in tabular form in 1880 (270), showed that protein decomposition in the cotyledons did not give anything like as high a proportion of asparagine as he found in the axial organs.

TABLE 13.

(From Schulze, 270.)

Etiolated seedlings of *Lupinus luteus*	In percentages of total N in the protein-free extracts of the cotyledons		In percentages of total N in the protein-free extracts of the axial organs	
	Aspara-gine-N	Other soluble-N	Aspara-gine-N	Other soluble-N
4 days old	17.5	82.5	70.0	30.0
6 " "	20.5	79.5	68.8	31.2
12 " "	26.2	73.8	78.1	21.9

In the extracts from both the cotyledons and axial organs, the "other soluble-N" consisted in large part of "amido-N" (Sachsse-Kormann) (262) and it would appear that these "amides," formed by protein decomposition in the cotyledons, were in large part converted into asparagine either as they streamed up into the axial organs or after their arrival therein. This would explain

* This experiment was made in December when the light was poor. The seedlings became green fairly soon, but new growth was very slow. The asparagine content was 29.3 per cent of the dry weight of the seedlings, 25.6 per cent being obtained from the sap by direct crystallisation without treating the mother-liquors with alcohol!

the very striking fact that the *concentration* of asparagine in the axial saps was always greater than that of the cotyledons.

TABLE 14.

(From Schulze, 270.)

Seedlings of *Lupinus luteus*	Per 100 g. of water	
	In the sap of the fresh cotyledons	In the sap of the fresh axial organs
	g.	g.
Grown 4 days in dark	0.65	1.30
Grown 6 days in dark	1.09	1.45
Grown 11 days in dark	1.23	1.62
Grown 12 days in dark	1.07	1.18
Grown 12 days in light	1.49	2.00

We cannot doubt (Schulze argued) that, in the plant, dissolved substances can be distributed by diffusion; and the above-mentioned relationships could surely not hold if the enormous accumulation of asparagine found in the axial organs were due to a continued and intensive production in the cotyledons. The transformation from other protein decomposition products therefore undoubtedly took place in the axial organs, and the question then arose, in what way was this brought about?

In 1878, Schulze could offer no simple chemical explanation, and very guardedly made the suggestion that it might be due to alternate protein decomposition and regeneration in the growing parts, the decomposition giving asparagine and other nitrogenous products and the regeneration taking place preferentially from the latter. By repetition of this process, the greater part of the cotyledon protein decomposition products would thus be ultimately transformed to asparagine, which would necessarily accumulate in the growing parts.

Such an admittedly speculative view, taken in conjunction with the previous suggestion that certain of the reserve protein decomposition products are preferentially utilised for protein synthesis in the axial organs, was, he felt, in keeping with the original postulate of von Gorup-Besanez, and explained why in some seedlings only small amounts of certain of the acid hydrolysis products of seed proteins could be found, while others showed a large accumulation of asparagine or glutamine.

In 1880, Schulze (270) again discussed very fully the question of asparagine formation in plants and stated that he now felt a little more confident about his views on the transformation of protein decomposition products to this substance. By identifying it with the activities of protoplasm, undoubtedly he had been very daring, for although Pfeffer (210) in 1876 had admitted that the processes of respiration might require a continuous breakdown and regeneration of protein, he was most emphatic that asparagine was to be regarded simply as a reserve protein material. Borodin's recent paper (20), however, had given him great encouragement, since he envisaged that, in the living protoplasm of all parts of all plants, a continuous interchange between protein and asparagine was essential for activities connected with respiration and growth of cells. Taken in conjunction with his own views given above—that the protein decomposition products suitable for protein regeneration vary from plant to plant—Borodin's suggestion might be used to explain the accumulation not only of asparagine or glutamine, but of other protein decomposition products in any part of the plant, and this would hold throughout the plant kingdom. His own investigations so far had been confined almost entirely to seedlings, but, following Borodin, he had looked for and found in twigs of *Alnus,* water-cultured in the dark for 10 days, asparagine and small amounts of "amido-N" in the etiolated leaf buds. Protein breakdown in leaves therefore probably gave the same mixture of products as in seedlings, and these might arise either directly, according to von Gorup-Besanez's postulate, or indirectly *via* protoplasm. In the green plant, however, asparagine might also be formed from sources other than protein. Kellner (116) had recently investigated various forage plants, and had found in the early stages much non-protein nitrogen, partly in the form of amide nitrogen (Sachsse), while Emmerling (76) had examined all parts of the broad bean (*Vicia Faba*) at various stages of growth, and had always found evidence for "amido-N" (Sachsse-Kormann). Kellner had suggested that, in the green plant, it was possible that asparagine was formed not only from decomposition of protein, but also from inorganic nitrogen translocated from the soil, and that it was difficult to decide which process was operative. Emmerling had agreed, and he (Schulze) felt that the latter alternative was not by any means improbable.

The above-mentioned views of Schulze on the mechanism of asparagine synthesis in seedlings were, of course, directly opposed to those of Pfeffer, and were not allowed to go unchallenged. In his *Pflanzenphysiologie,* published in 1880, Pfeffer (211) stated that Schulze's assumption that the breakdown of protein in plants must give rise to amino acids and acid amides in the same proportions as under the action of known chemical agents could not be accepted without further enquiry, for already numerous *chemical* experiments could be cited* which indicated that the decomposition products varied according to the treatment to which the proteins were subjected. The plant evidently possessed the ability to build up protein, which contained many molecular complexes, from any one of a number of simple nutrient materials, so it could not be

* The contemporary literature contained many data which supported Pfeffer's statement. Ritthausen, for instance, isolated 2.0 per cent of aspartic acid from conglutin after acid hydrolysis, Hlasiwetz and Habermann (109), 14.5 per cent of aspartic and "malaminic" acid (the latter being considered an uncrystallisable isomer of the former), after treatment of the same protein with bromine. Schützenberger (303) observed that alkaline hydrolysis of albumin gave a mixture of amino acids from which he separated valine, then not yet recognised as a product of acid hydrolysis. There was also Nasse's "firmly bound" ammonia, given only on prolonged alkaline hydrolysis. These contradictions are readily explainable today, but to Pfeffer and certain of his contemporaries they were naturally legitimate ground for argument.

denied that on decomposition these complexes might be broken down in a manner specific to the particular organism.

In 1885, Schulze (271) replied that the chief task of physiological chemistry was to make possible an understanding of the changes taking place in the organism by identifying them with *chemical processes* that can be brought about outside it, and all that he had attempted to do in this particular case was to bring observed plant phenomena into harmony with the results obtained by decomposing proteins with known chemical agents. He reminded Pfeffer that, since the time when Lavoisier gave chemists the balance, chemical change had been studied quantitatively as well as qualitatively! von Gorup-Besanez's postulate could not explain *quantitatively* the chemical changes observed during the germination of lupins, etc., and other explanations were needed. Pfeffer's own suggestion, however, was in his opinion very unlikely, for there was much chemical evidence (summarised in 1882 by Baumann [9]) to show that certain atomic groupings present in protein decomposition products pre-existed in the protein molecule. Proteins, for instance, gave the Millon test for tyrosine, suggesting that a phenolic-hydroxyl group was present, while oxidation gave benzoic acid and benzaldehyde in sufficient amount to warrant the assumption that another aromatic group was also present (Schulze suggested that this might be due to phenylalanine). It was true that Löw (139) in 1875 had doubted the presence of a leucine complex because the free amino acid slowly reduced osmic acid, while proteins themselves did not, but Baumann had pointed out that the comparison was not a valid one since glucose and fructose both reduce Fehling's solution, while cane sugar does not. Moreover, in a later paper, Löw (141) had admitted that both aromatic and phenolic nuclei might be present in the intact molecule even though he still doubted whether the leucine complex was preformed, since he could not obtain valeric acid on oxidation.

I have pointed out earlier on page 12 the vague ideas on the constitution of proteins current in 1872 when Pfeffer was formulating his ideas on asparagine in seedlings; since then much progress had been made, but only with respect to the nature of the protein decomposition products; there was still no evidence which would help chemists to picture in their minds the actual constitution of the protein molecule itself. It is not surprising therefore that individual notions on this important question were rampant and that chemists should disagree among themselves no less than with plant physiologists. Pflüger's assumption* that "living protoplasm" was endowed with unique chemical activities was, for instance, strongly supported by Detmer (66), who claimed that in the plant it brought about "dissociation" of protein into nitrogenous and non-nitrogenous substances, instancing the elaboration of fat from protein in animals. In the plant, breakdown of protein gave "amides" (asparagine, leucine, etc.) and carbohydrate, as Pfeffer had observed, but this was due to the wonderful union (*wunderbare Verknüpfung*) of the atoms in the living cell, and it could not occur outside the plant under the action of chemical agents because the protein had, by then, been stabilised.

Small wonder that, during this period (of "Vitalism"), Schulze himself should have been influenced by the prevailing confusion, and have been very cautious in his criticisms! If amino acid complexes were preformed in the protein molecule, as he firmly believed, then Pfeffer's reiterated postulate, that the whole of the asparagine found in lupin seedlings came directly from the seed protein conglutin, was impossible, since his assistant, Bosshard, had found that this protein gave on acid hydrolysis 13 per cent of its nitrogen as "loosely bound nitrogen" (what we now call "amide nitrogen") and the maximum amount

* In 1875, Pflüger (213) had suggested that the protein of "dead" protoplasm was stabilised and behaved as chemists observed, but that "living" protoplasm was continually undergoing intermolecular change, the amino group being transformed to the nitrile.

of primary asparagine and glutamine that could be present in the seedlings therefore was equivalent to 26 per cent of the total nitrogen of the decomposed protein. The origin of the excess asparagine was admittedly speculative, but he preferred his own view that it was due to the continuous decomposition and regeneration of protein, as this could be considered in keeping with the known facts of protein chemistry. "Living" protoplasm was a possibility, but he had no faith in it!

4. *Schulze Modifies His Views on the Origin of Asparagine in Seedlings.*

A perusal of Schulze's papers published during the 10 years previous to this verbal duel with Pfeffer in 1885 suggests that he was, at heart, not really at all favourably disposed towards any explanation of metabolism which necessitated the aid of "protoplasmic activity." The accumulation of asparagine in plants was a chemical phenomenon and he realised all along that explanations were difficult because no clear idea of even the rudiments of protein constitution had been formulated, and so very little was known about the numerous other non-protein nitrogenous constituents of plant organs and their metabolism. His own researches were directed towards an improvement in the latter field, and for this reason he undoubtedly regarded his discovery with Barbieri (288) of phenylalanine in lupin seedlings, and the proof a little later that it was a constituent of the lupin protein conglutin, as of great importance, since such evidence emphasised the close connection between amino acids and protein metabolism. The same was true of their isolation in 1883 of valine from lupin seedlings, an amino acid then known only as a constituent of pancreas juice. Continuing his studies with lupin seedlings, he was able, in 1886, to announce with Steiger (297, 298) the discovery of a new base which they called arginine. This was present also in squash seedlings, and a study of its properties showed *inter alia* that it gave ammonia and carbon dioxide on being boiled with alkalies. Five years later Schulze (273) found that the amount of arginine produced in lupin seedlings exceeded the total non-protein nitrogen originally present in the seed; it must therefore have been derived from the seed reserve proteins. The eventual proof by Hedin that it was a true acid hydrolysis product of proteins, and that it was a component of Drechsel's "lysatin," makes fascinating reading (cf. Vickery and Schmidt, 339) and illustrates the important bearing that Schulze's plant researches had at that time on protein chemistry. Arginine, of course, explained Nasse's "firmly bound" ammonia.

Meanwhile, in 1888, Schulze was forced back into polemics by the publication the year before of a stimulating but provocative paper by Müller (181), who called to question his views on asparagine metabolism. Müller's work was on similar lines to that of Borodin, whose microchemical technique he followed, and it concerned the relationship between asparagine and protein in leaves. His conclusions were highly speculative and were based in part on the original view of Pfeffer, restated more recently by Emmerling (76), that asparagine and carbohydrate could condense *en bloc* to give protein (protoplasm). Briefly, he developed the thesis that it was the assimilation process itself—with the carbohydrate *in statu nascendi*—which led to the transformation of asparagine to protein, and, in a discussion on the origin of the asparagine, he wrote:

Whether in seedlings the amide in question arises from the reserve proteins I will leave undetermined. But it must be mentioned that the quantity found far exceeds that which is produced from protein by artificial decomposition. Hence the production of asparagine from protein according to Borodin seems to me to contradict the principle of "minimum of work" [*Arbeitersparniss*]. Why should protoplasm when once formed decompose again only to be regenerated? It is much more probable that the asparagine in the plant arises from the assimilated carbohydrate and the inorganic nitrogen compounds [he had stated earlier that this was Emmerling's view] and that later, through the assimilation process, the asparagine is further transformed to protein. . . . It is taught that the fermentation of sugar and of malic acid produces succinic acid. Fermentable sugar and malic acid are widely distributed in plants, and processes that resemble fermentation, namely, respiration, readily go on in the plant. Through this process carbohydrate can be transformed into alcohol and, I suspect, also into succinic acid just as in the case of fermentation. In the presence of inorganic nitrogen, therefore, the possibility of asparagine synthesis from succinic acid cannot be disregarded. Further, it may arise in the same way direct from the malic acid in the plant,* and, in this case also, carbohydrate must be regarded as necessary for its formation, since this acid owes its origin to it. In both cases, accordingly, carbohydrates are required for asparagine enrichment.

Schulze's (272) paper of the following year (1888) was, in part, a reply to Müller and, in part, a long discussion in which he evolved a new hypothesis of asparagine production from proteins which has stood the test of time. The paper portrays him, if I may be permitted the expression, in his most lovable mood. First of all, he deals justly with Müller's ideas about carbohydrate *in statu nascendi,* which his own evidence will not permit him to accept; next, waxes mildly indignant over the suggestion that there can be any doubt whatever about the origin of asparagine in seedlings; then admits, after some prevarication, the force of Müller's ridicule of Borodin and, finally, puts forth in outline the new theory mentioned above, which, he states, has so little chemical evidence behind it that he will not trouble to discuss it further—and then promptly does so, to the extent of nearly 2000 words!

With regard to asparagine production in green plants, Schulze pointed out that Müller had misquoted Emmerling (77), who maintained that it might arise not only from inorganic nitrogen and carbohydrate, but also from

* Müller seems to have been unaware that Beyer suggested this in 1867! (Cf. p. 6.)

regressive protein metabolism; Kellner had expressed similar views in 1879, and he (Schulze) had agreed with them (cf. p. 42). The nitrate and ammonia content of mature green plants was, however, so very small that he felt the second alternative was the more usual process employed, especially in the case when plants were kept in the dark. There was no definite evidence on the point, but he thought it likely that protein breakdown in leaves might give the same mixture of products (asparagine, glutamine, "amides," etc.) as in seedlings; on the other hand, it was possible that the protein molecule itself, or perhaps the above-mentioned decomposition products, might give rise to carbon-free or carbon-poor nitrogen-containing residues* and that these could be used, with nitrogen-free organic compounds, to synthesise asparagine.

In the case of seedlings, Schulze was emphatic that his earlier work with lupins left no doubt whatever that the asparagine came from the seed reserve proteins, for the 15-day etiolated seedlings contained 4 per cent of asparagine nitrogen and the total non-protein nitrogen present in the ungerminated seed was only 0.46 per cent. These figures showed that protein must have furnished the major part of the material for asparagine building, and this would hold whether one regarded the latter as being elaborated according to his postulate of 1878, or the one just propounded, involving nitrogen residues and nitrogen-free organic material. Müller's criticism of the point was not very precisely phrased, and Schulze no doubt thought that he had answered it in a satisfactory manner —actually, however, he was slightly confusing the issue, for all that he had, in fact, shown was that the asparagine nitrogen must have originated, in very large part at least,

* A footnote to this statement reads as follows: "Whether it is possible that by such a decomposition process ammonia residues may be formed or some others which contain nitrogen-rich groups is a question upon which so little that is positive is known that I shall not trouble to discuss it further here." Actually, he does this, a few pages on, at great length, but is far too cautious to mention ammonia again specifically!

from protein. He did not discuss the origin of the carbon skeleton of the asparagine or of the nitrogen-free organic material at all, and possibly took it for granted that these came also from the protein. If, however, the protein nitrogen became completely detached from its original protein complex and reacted in the form of "nitrogen residues," then the nitrogen-free organic material with which these condensed might quite likely have come from some other, independent, source. In a vague way, for he was no chemist, this is perhaps what Müller wished to imply. The question is an important one in connection with the possible respiratory activities of proteins in plants, and I shall return to it later.

Schulze's new idea—undoubtedly prompted by Müller's views on asparagine formation—that protein breakdown in plants might go beyond the amino acid stage and give nitrogen-rich residues seemed to him at first such a wild speculation that he immediately hedged by stating that he still had great faith in his former hypothesis that asparagine was formed from amino acids *via* alternate regeneration and decomposition of protein! Borodin had been loth to believe that proteins were not concerned in the active changes associated with protoplasm, and Schulze still agreed with him. He admitted, however, that others, besides Müller, were prepared to question this; Emmerling (77), for instance, asserted that in plants the proteins, when once built up into protoplasm, were used again only very sparingly in metabolic processes.

Fortunately the quaint mental kink which bid Schulze shelve a new idea because it was too speculative for his cautious nature was not dominant for long; the degradation of amino acids, although shrouded at the time in mystery, was a process that might one day yield to chemical investigation, and the chemist in him could not remain silent! Returning to the question, he put forward in a very guarded way what he called his new hypothesis. Direct protein decomposition in the plant produced asparagine, glutamine, leucine, etc., as von Gorup-Besanez had originally suggested, and in the proportions (with the amides as acids + ammonia) given on acid hydrolysis.* The greater part of these products then underwent further decomposition to give carbon-free or carbon-poor nitrogenous residues (he had not the courage to mention ammonia this time!) which could be used with nitrogen-free substances in synthetic processes to build up asparagine. Nitrogen-free raw material for this purpose was, as a rule, present in plants, and he agreed with Müller that it could give malic and succinic acids on oxidation, which stand in a known close relationship to asparagine.†

The small amounts of leucine, tyrosine, phenylalanine, etc., found in seedlings, perhaps also the glutamine and part of the asparagine can thus be regarded as products of *primary* protein decomposition; the remainder of the leucine, etc., which has disappeared, has undergone a degradation to give nitrogen-rich residues, and these have condensed with organic acids to build up asparagine, which must therefore be in part of *secondary origin*.‡ The new hypothesis was attractive because it made no use of the contentious idea of alternate protein decomposition and regeneration, but Schulze was not willing to bury the

* von Gorup-Besanez's announcement (p. 35) that he had extracted from vetch seedlings a ferment which would produce peptone from albumin was questioned by Krauch (122), who maintained that the extract itself contained peptone. The controversy was cleared up by Green (96), who pointed out the inadequacy of the latter's experimental work, and showed that, in the seeds of *Lupinus hirsutus,* there was a ferment, present as zymogen, which was activated in very weak acid solution, and would then digest fibrin to peptone, leucine and tyrosine. Green was at the time an assistant to Vines at Cambridge, who in later years (1902–1910) showed that peptases and ereptases were widespread in plants (340).

† Schulze here definitely suggests that the carbon skeleton of asparagine might come from nitrogen-free raw material (carbohydrate); see my discussion earlier on p. 49.

‡ Towards the end of his paper he briefly discussed the possibility that *all* the organic nitrogen compounds in plants are products of synthesis from carbon-free or carbon-poor products derived from protein decomposition. According to this view *all* the asparagine and amino acids would be of *secondary origin*.

ERNST SCHULZE

old one without an emphatic statement to the effect that there was no evidence whatever to show that in plants amino acid degradation of the type he was now suggesting could take place at all! Müller's "carbohydrate *in statu nascendi*" may have been merely the immature idea of a young man anxious to impress the plant physiological world, but at least he had the merit of having forced Schulze to reconsider a weak position and put forward suggestions as to amino acid metabolism which were really 15 years ahead of their time.

It was not until 1906 (285), following the first suggestion that amino acids could undergo deamination with the production of ammonia and the corresponding α-ketonic acid, that Schulze would definitely commit himself to the view that ammonia was the "nitrogen-rich residue" referred to in the above hypothesis. Meanwhile, in 1892 (274), he discussed briefly Löw's (140) suggestion that there were two types of protein breakdown in plants, (i) a tryptic fermentation to give amino acids, etc., and (ii) an oxidation of protein by "living" protoplasm to give ammonia; one protein molecule giving aspartic acid as per (i) and another giving ammonia as per (ii), the two products uniting with loss of water to give asparagine. The idea that the production of asparagine from protein was an oxidation goes back of course to Pfeffer (208), and Sachsse (261) in 1876 had constructed an equation to represent the reaction. It had been revived again in 1888 by Palladin (201), whose results (deduced from experiments with plants deprived of oxygen, which showed no accumulation of asparagine) Löw regarded as confirmation of his views. Clausen (53), however, had objected that lack of oxygen killed plants in 24 hours, and that the changes observed by Palladin were really *post mortem*. Schulze (274) reviewed this evidence and was not convinced; in his opinion no significant change in either protein or asparagine could be observed in such a short interval as 24 hours! With regard to Löw's dual mechanism for asparagine production he agreed that it might, if modified in certain details, be used to explain his observation, but pointed out that there was no evidence for protein oxidation as in (ii). If oxidation occurred, he thought that it would be at a later stage, in the conversion of amino acids, especially of the fatty series, to asparagine since he had never been able to detect alanine or glycine in seedling saps. This suggestion he repeated in 1898 (280), but would not finally adopt it until 1906, as stated above, although his brilliant pupil Prianischnikow used it as a working hypothesis from 1899 onwards.

I need mention only one other point before leaving the historical approach to protein metabolism in seedlings. In 1898, Schulze (280) remarked that for 20 years he had considered that the amino acids, etc., resulting from the decomposition of the reserve proteins in the cotyledons were transported to the axial organs before conversion into asparagine and that this was what had induced him to suggest first his old and later his new hypothesis of asparagine formation. On the basis of the then known facts, he had held that leucine, tyrosine, etc., were more suitable for protein synthesis than asparagine, which was the reason why the asparagine accumulated in the regions where it was formed. The new mechanism for asparagine synthesis, however, led him to adopt now the opposite view and to regard asparagine (or glutamine) as *more easily available* for protein synthesis than the amino acids. Asparagine formation was thus the *first phase of protein regeneration*. This change-over was to bring him a little later into conflict with Prianischnikow.

Here we can take our leave of Schulze's controversy respecting the origin of asparagine in seedlings. His new hypothesis was definitely accepted, although the final proof came from the work of a later generation. From 1900 onwards, he was to witness the important advances in protein composition and structure made by Emil Fischer—final proof, in fact, that all the amino acids were indeed preformed in the protein molecule—and the early work on amino acid metabolism carried out by Neubauer,

Embden and Knoop. But by then he had become an old man and his eyesight was failing—henceforth he left speculation to others, and contented himself in his declining years with a pastime in which he has had no peer in any country, the isolation of amino acids and other nitrogenous products from plant organs at various stages of physiological activity.

CHAPTER III

PROTEIN METABOLISM IN SEEDLINGS REVIEW OF THE LATER LITERATURE

1. *Introduction.*

WE saw in the last chapter that Schulze, after many vicissitudes, had arrived at the general conclusion that, on germination, the reserve proteins of the seed are broken down by enzymes to give peptones and amino acids, and that these latter undergo some further change to give nitrogen rich residues, possibly ammonia, which then in turn condense with organic acids to give asparagine, or, in certain species, glutamine. The amide thus formed accumulates in the growing parts to be used preferentially for protein synthesis *in situ*. Moreover, the extent of protein breakdown in the cotyledon or endosperm, of new synthesis in the growing parts and of amide accumulation are functions of the available carbohydrate, probably glucose.

These views he amplified considerably in his later years, and it is to him and to Prianischnikow that we owe most of our present knowledge of seedling metabolism; for, since their work, surprisingly few investigations have been undertaken with seedlings, the focus of interest, especially in recent years, having shifted to the leaves of green plants, organs which these earlier workers found too complex for the chemical methods then available. Notwithstanding this, many observations on seedling metabolism, which were confusing in those days, now receive what appears to be a reasonable explanation in the light of advances made in cognate fields of research, as we shall see later.

2. *Asparagine as a Secondary Product of Seedling Metabolism.*

I have already emphasised (p. 45) the importance that Schulze attached to the isolation and identification of the non-protein nitrogenous products of seedlings, for he regarded them as the key to the protein metabolism. In the last general summary of his work (285) published in 1906, he was able to report the presence of ten of the amino acids known to be hydrolysis products of protein, viz. valine, leucine, isoleucine, phenylalanine, tyrosine, tryptophan, proline, arginine, histidine and lysine. These could be obtained in appreciable quantity only after germination, and were therefore undoubtedly derived from the seed reserve proteins.* He had, of course, isolated asparagine and glutamine, but at the time he could only surmise that these might pre-exist (in part) in the protein molecule, for the direct proof of this possibility dates from recent times (59, 60). To the best of my knowledge, this list is nearly as complete as we can make it today; Shorey (308) in 1897 had obtained evidence for the presence of glycine in sugar cane, while Vickery (331) has since found alanine and serine in the green parts of *Medicago sativa*.

In keeping with his view that the elaboration of asparagine in seedlings took place in the growing parts from amino acids which had streamed up from the cotyledons, he was able to isolate these amino acids in larger amount and in greater variety from the latter organs. Some of his

* Schulze had, in addition, isolated in small amount many other nitrogenous products from seeds or seedlings; e.g. allantoin, xanthine, etc., which he thought might arise from breakdown of nucleoprotein; betaines such as trigonelline and stachydrine, and alkaloids such as ricinin which he thought might be synthesised from amino acids derived from the reserve proteins. Such products, however, can play either directly or indirectly only a minor role in the general protein metabolism of the plant and I do not discuss them further in this book. An excellent review of the more recent work in these fields has recently been published by McKee (158).

results with Castoro (294), published in 1903, and of his pupil Wassilieff (353), using white lupins (*Lupinus albus*) growing in the light illustrate this point.

The nitrogen partition at various stages of growth was as follows:

TABLE 15.

(Compiled from Schulze and Castoro, 294, and Wassilieff, 353.)

Lupinus albus germinated in light	Seeds without testas	Per 100 g. of ungerminated seeds without testas		
		4-day seedlings	7-day seedlings	14-day seedlings
	g.	g.	g.	g.
Protein-N	6.89	5.38	3.26	3.34
Asparagine-N	–	1.21	2.47	3.06
Arginine-N	0.007	0.08	0.04	0.01
Other-N (diff.)	0.78	1.01	1.91	1.27
Total-N	7.68	7.68	7.68	7.68

At the 4-day stage, when the hypocotyl was 1.5 cm. long, he found asparagine, leucine, tyrosine, arginine, histidine and lysine in the cotyledons, but only asparagine, leucine, valine and phenylalanine in the growing parts. The protein nitrogen had decreased, and the major part of the deficit had already reappeared as asparagine nitrogen. This change was even more marked at the 7- and 14-day stages and, in the latter case, the asparagine nitrogen represented 86.5 per cent of the lost protein nitrogen—an amazing result! In spite of this, however, arginine, histidine and leucine could still be isolated from the cotyledons, and traces of the two bases were found in both the stems and leaflets. As very little synthesis of new protein had then taken place, these results leave no doubt that the amino acids had been metabolised to provide nitrogen for asparagine and entirely confirmed his original suggestion of 1888 (p. 47).

It might be thought that Schulze was rather labouring

the point here because his much earlier results of 1878 (table 6, p. 20) show that "amido-N" had contributed to the increase in the asparagine nitrogen observed. Accepting his change of viewpoint regarding the origin of the asparagine, this is indeed true, but it must be remembered that in those days he and Umlauft had isolated only leucine and tyrosine from the seedling sap, and until he had proved to his own satisfaction that the total amount of amino acid present was comparable to that suggested by the "amido-N," he does not appear to have placed much reliance on the latter value. Throughout his whole life he was, in fact, distrustful of all indirect methods of analysis and the only one he used consistently—that of Sachsse for determining amide nitrogen—he frequently checked by a direct isolation, and was not content unless he obtained over 80 per cent of the amount expected; an attitude in striking contrast to that displayed in some of the modern work with leaves!

Some five years before Schulze's long and laborious work was published, Prianischnikow (225) had actually shown very neatly that part of the asparagine in seedlings must come from protein decomposition products. Using some results of his own, and of Schulze's pupil, Merlis (164), he compared the *rate* of protein decomposition and the *rate* of asparagine accumulation (see fig. 1, p. 77). Both reached a maximum at about the same time, but later the former decreased much more rapidly than the latter, showing that asparagine was then being synthesised in part from other products of protein decomposition, and Prianischnikow thought it probable that these must be the amino acids.

In the year 1899, this suggestion was speculative, and Schulze undoubtedly regarded it as such, for at that time protein analyses were very far from complete (*circa* 45 per cent), and there was still no evidence that, on enzymic or even on acid hydrolysis, the protein molecule would give only amino acids. The contrast between the attitude of master and pupil, between the experience of ripe years and the rashness, though possibly keener intuition, of youth, appears very vividly here; Schulze, as we have seen, spent five laborious years definitely proving a point that Prianischnikow was content to infer from the earlier work of Schulze's own school!

3. Glutamine as a Secondary Product of Seedling Metabolism.

In many seedlings, particularly in plants of the orders *Polypodiaceae, Polygonaceae, Chenopodiaceae, Caryophyllaceae, Umbelliferae* and *Cruciferae,* Schulze found that the protein metabolism led to an accumulation of glutamine instead of asparagine. The distinction, however, was not sharp, and appeared to depend on climatic conditions and age of the plant: sometimes one amide was found, sometimes the other, and in numerous cases both amides were actually isolated (cf. Stieger, 316). In general, he found that glutamine was elaborated on germination of oil-bearing seeds, and as these contained relatively little protein and abundant reserve nitrogen-free material for growth and respiration, the breakdown of protein was not intense. The amount of glutamine formed, therefore, was small compared with that of asparagine, and the maximum amount he recorded was 2.5 per cent (of the total dry weight) isolated from the axial organs of etiolated castor bean (*Ricinus communis*) seedlings (279). Schulze never determined glutamine nitrogen indirectly by the method of Sachsse (mild hydrolysis followed by estimation of liberated NH_3), as he could never be certain that asparagine was absent; Prianischnikow (224), however, records 6.17 per cent (in terms of total seedling nitrogen) in seedlings of squash (*Cucurbita pepo*). That Schulze's caution was fully justified is shown by some recent work of Schwab (304), who has examined many seedlings and leaves by a modification of the Chibnall and Westall (52) method for determining glutamine in the presence of asparagine. A selection of his results for seedlings is given below.

TABLE 16.
(From Schwab, 304.)

Seedlings of	In percentages of total-N	
	Glutamine N	Asparagine N
Helianthus annuus	3.47	3.63
Ricinus communis	4.14	1.69
Cucurbita pepo	2.10	2.21
Linum usitatissimum	1.62	1.68
Phaseolus multiflorus	3.2	22.4
Lupinus luteus	1.0	20.0

Schulze's assumption that glutamine was associated with fat metabolism is clearly incorrect, for the oil-bearing seeds on germination all gave reasonable amounts of asparagine, especially *Cucurbita pepo*. Schwab also found both amides in many *Gramineae* seedlings (confirming Schulze), but I am inclined to doubt if the method of glutamine analysis is sufficiently accurate to establish the validity of the small value quoted for the yellow lupin in table 16, even though the production of primary glutamine at this stage is probable. Schwab's results, however, leave no doubt that, in most families, both amides may be elaborated, and in proportions which vary in the different regions and with the stage of development of the plant.

I shall give my own views on the comparative roles of the two amides later, but I can quote here some unpublished results obtained in my laboratory by Westall showing the nitrogen partition during the germination of *Ricinus communis,* glutamine nitrogen being separately estimated by the above-mentioned method.

The seeds contained a small amount of asparagine which was slowly replaced by glutamine as the seedling developed. Appropriate analyses of the hypocotyls and roots at the 8- and 15-day stages showed that they contained the major part of the glutamine (confirming Schulze, 279) and no asparagine, the latter being confined to the cotyledons. It is interesting to note that Green

TABLE 17.

(Unpublished data.)

Per 100 g. of ungerminated seeds

Ricinus communis planted 24.iv.1933	Seeds	Seedlings 6 days in light	Seedlings 8 days in light	Seedlings 8 days in light followed by 7 days in dark	Seedlings 8 days in light followed by 14 days in dark
	g.	g.	g.	g.	g.
Protein-N	2.34	2.19	1.96	1.54	1.65
Glutamine-N	0.004	0.085	0.14	0.30	0.44
Asparagine-N	0.141	0.065	0.03	0.02	–
Other soluble-N (diff.)	0.502	0.655	0.87	1.14	0.89
Total-N	3.0	3.0	3.0	3.0	3.0

(97) isolated a small amount of asparagine from very young seedlings of this plant, which shows that the indirect methods of analysis are valid.

As the amount of glutamine elaborated in seedlings appears to be so much less than that of asparagine, in e.g. leguminous seedlings, it is pertinent to enquire whether this amide might not be regarded simply as a direct primary protein decomposition product, as Nuccorini (188) has recently claimed.

Schulze refrained from making a positive pronouncement on this point, for although he regarded the primary production of both amides as possible, he had no proof, and, since the nitrogen of asparagine had clearly to arise in large part by some secondary change involving the amino acids, he thought the same probably held for that of glutamine. We know, however, from the recent work of Damodaran (59) that glutamine pre-exists in wheat gliadin, and of Damodaran, Jaaback and Chibnall (60) that asparagine pre-exists in edestin, consequently we are now justified in assigning all the so-called "amide-N" to pre-existing amides in the protein molecule, and in regarding the amount of glutamic acid obtained after acid hydrolysis as a measure of the *maximum* amount of glutamine which that protein can give on enzymic digestion. Fortunately the two main reserve proteins of *Ricinus* seeds, a crystalline globulin and an albumin, ricin, have been analysed, so that the maximum amount of primary glutamine that these can give on germination of the seed is known. Relevant data are collected in table 18. In Westall's experiment, the 14-day seedling growth produced 2.3 per cent of glutamine concomitant with a fall in total protein of 5 per cent, both values being calculated in terms of total dry weight of ungerminated seed. It will be seen that the amount of glutamine produced is far in excess of that which could arise from the primary decomposition of the two reserve seed proteins, even allowing for possible inadequacies in the analysis for glutamic acid, and assuming that the whole of this acid pre-existed as its amide.

TABLE 18.

Ricinus communis	Ungerminated seed		14-day seedling
	Globulin	Ricin	
	% of protein	% of protein	% of decomposed protein
Glutamic acid	14.5*	20.0†	–
Glutamine	–	–	46

From these results I can draw only one conclusion, namely, that as seed proteins contain no inconsiderable proportion of glutamine, the production of a limited amount of this amide during seedling metabolism through primary protein decomposition is possible, but that a secondary origin involving *inter alia* amino acids can also occur. The contrary suggestion of Nuccorini was clearly based on his own (low) yields of crystalline glutamine from the etiolated seedlings.

Both asparagine and glutamine, therefore, must be products which can have a secondary origin in seedling

* Osborne and Gilbert (194).
† Karrer *et al.* (114).

metabolism, but before passing on to discuss the more general question of protein decomposition and regeneration in seedlings, I should like to mention the suggestion of Suzuki (320) that, in coniferous plants, arginine can also function as a product of secondary metabolism, since it will permit me to introduce one of the most beautiful examples of analytical chemistry performed by Schulze and his colleagues.

4. *Arginine as a Primary Product of Protein Decomposition in Seedlings.*

Suzuki claimed, from very inadequate analyses of seedlings of *Pinus Thunbergii,* that arginine was formed at the expense of other amino acids, that it had the same physiological function as asparagine, and that it was synthesised in preference to the latter when ammonia was fed to the seedling. The fact that arginine sometimes accumulates in seedlings, especially in conifers, was already well known to Schulze (278). With certain plants, viz. the white lupin (*Lupinus albus*), the blue lupin (*Lupinus angustifolius*), the vetch (*Vicia sativa*), and the pea (*Pisum sativum*), he could find only traces of arginine in the seedlings, and concluded that this amino acid must be utilised, with the other protein decomposition products, for the formation of new protein or asparagine. But in etiolated seedlings of the yellow lupin (*Lupinus luteus*), there was always present a high and fairly constant amount of material, including arginine, precipitable by phosphotungstic acid, and this in spite of a large elaboration of asparagine.

To investigate this abnormality, he determined (with Winterstein, 300) the arginine content of the total protein of lupin seeds by the usual Kossel-Patten procedure, and then (with Castoro, 295) the arginine content of the seedling sap at various stages of growth, the same procedure being applied after a preliminary treatment necessitated by the complex nature of the other products present. The data given below show that the amount of

TABLE 19.

(From Schulze, 279.)

Lupinus luteus   Per 100 g. of ungerminated seeds without testas

	Ungerminated seeds without testas	6-day etiolated seedlings	15-day etiolated seedlings	Seedlings grown 14 days in dark and then 10 days in light
	g.	g.	g.	g.
Total weight	100	95.02	77.20	82.22
Protein-N	8.72	5.49	1.71	1.78
Asparagine-N	–	1.16	4.02	5.09
Basic-N	0.46	0.97	1.22	1.03
Rest-N	0.16	1.72	2.39	1.40
Total-N	9.34	9.34	9.34	9.34

arginine in the sap was never in excess of that which must have been formed by primary protein decomposition.

Per 100 g. of decomposed protein:
The 6-day seedlings contained 6.31 g. arginine
The 11-day seedlings contained 6.32 g. arginine
The 15–16-day seedlings contained 6.70 g. arginine
100 g. seed protein contained 6.9 g. arginine

Schulze stated specifically that the arginine values for the seedling saps were to be regarded only as minimal, and that he quoted the above-mentioned value for the seed protein itself only because it was determined under comparable conditions. A second analysis which he made, with Castoro, of the total protein of lupin seeds (prepared by Ritthausen's method of extraction with dilute aqueous soda) gave 7.86 per cent arginine, while a recent analysis of the total extracted globulin made in my laboratory by Dr. Tristram gave 11 per cent arginine, so that the true balance was probably not as close as that suggested by the figures given above, and sufficient excess of arginine was no doubt available for protein regeneration or even for secondary conversion to provide nitrogen for asparagine.

Suzuki's claim with regard to conifers has been reinvestigated much more recently by Mothes (177). He severely criticised Suzuki's method of analysis, and has carried out a wide series of experiments with *Picea excelsa, Pinus Thunbergii, Pinus nigra, Pinus pinea* and *Abies Nordmanniana*. Total bases in the seed proteins and seedling saps were obtained by precipitating with phosphotungstic acid, and arginine by precipitation direct as the flavianate. Since the content of arginine nitrogen was found in a few cases to be about two-thirds that of the total basic nitrogen, this ratio was assumed to hold in the others also, an unwise procedure in my opinion, as Vickery (332) has shown that, in the sap of more mature plants, much of the basic nitrogen is of unknown composition. Mothes found, however, that the basic nitrogen of the newly formed protein was less than that preformed in the seeds, that the total basic nitrogen of the whole seedling did not change appreciably, and that therefore the content of free base in the sap depends on the rate of reutilisation for protein synthesis. His results varied somewhat from plant to plant, but in no case could arginine synthesis be demonstrated, and in certain cases, i.e. with *Picea excelsa*, the secondary conversion of base to asparagine was clearly shown. Furthermore, feeding the seedlings with ammonium salts gave the usual increase in asparagine; Mothes concluded therefore that, in conifers, arginine is a valuable and specific reserve substance that can be further metabolised if conditions demand. Klein and Tauböck (118a) report similar results with etiolated seedlings of *Pinus pinea*, which showed little or no utilisation of arginine in spite of the fact that nearly the whole amount of this base present in the seed (initially 94% combined in protein) was rapidly liberated in free form. Moreover the normal seedlings, studied over a period of 60 days, showed no evidence of arginine synthesis.

5. *The Decomposition of the Reserve Proteins on Germination.*

It will be recalled that Hartig (p. 6) in 1858 was the first to suggest that, on germination, the reserve proteins of the seed were broken down to provide soluble nourishment—"Gleis"—for the growing parts, and that this view was accepted later by Pfeffer, who pointed out that, in so far as nitrogenous material was concerned, "Gleis" was asparagine. Following von Gorup-Besanez's discovery in 1874 of proteolytic enzymes in vetch seeds, there was general agreement that the breakdown on germination was due to these agents, but Pfeffer's hypothesis, which held the field at this time, so stressed the importance of asparagine in protein metabolism that the factors which control the actual protein breakdown itself, as distinct from what was regarded as a concomitant elaboration of asparagine, do not appear to have been considered.

Schulze first discussed this question in his paper of 1880 (270). Accepting Borodin's modification of Pfeffer's hypothesis, which related protein regeneration from asparagine to the amount of physiologically active carbohydrates (glucose) present, he was able to show that the extent of protein breakdown during seedling growth was controlled in part by the quantity of reserve carbohydrate or fat present (cf. tables 10, 11, p. 33). He recalled, however, an observation to which he had already drawn attention in 1876, namely, that in the *initial stages* of germination rapid breakdown of protein occurred even when much carbohydrate was still present (cf. p. 17 and table 3). This puzzled Schulze, who returned to the question repeatedly in his later papers. In 1898, for instance, when he reviewed (279, 280) at great length the position of carbohydrate in protein regeneration from the standpoint of his new ideas on asparagine synthesis, he could still provide no satisfactory explanation. He was able to quote many new experiments showing that in the *later*

66 PROTEIN METABOLISM IN THE PLANT

stages of seedling growth (either in the dark or, if in the light, before extensive leaf development) the extent of protein breakdown, and incidently of asparagine accumulation, was conditioned by the nutritive (carbohydrate or fat to protein) ratio. Yet in the *early* stages of growth this could not apply, and he quoted the behaviour of yellow and white lupins, which contain the same reserve protein and nitrogen-free material but in very unlike ratios. After six days' growth in the dark, the extent of protein breakdown was almost the same in both cases.

TABLE 20.
(From Schulze, 279.)

	Per 100 g. of ungerminated seeds without testas	
	Lupinus luteus	Lupinus angustifolius
	g.	g.
Protein	52	37
N-free reserves (carbohydrate and fat)	24	46

Since, in the early stages of germination, physiologically active carbohydrate would never be lacking, he was forced to conclude that such substances could not protect the proteins from decomposition.

This question had also been discussed from a slightly different point of view three years before by Prianischnikow (222, 223), in a summary of the results of work carried out in Schulze's laboratory. Vetch seeds, which contained a higher nutritive (nitrogen-free reserve: protein) ratio (1.5:1) than lupin seeds (0.45:1), had been germinated in the dark and complete analyses made at regular intervals.

The comparison with Schulze's results for the lupin (table 12) is instructive, for the vetch seeds contained abundant starch and this, on germination, was converted in part (so Prianischnikow states) to sucrose. In agreement with Schulze's findings, the breakdown of protein in

TABLE 21.
(From Prianischnikow, 223.)

Seeds of *Vicia sativa* germinated in the dark — Per 100 g. of ungerminated seeds

Number of days' growth	0	10	20	30	40
	g.	g.	g.	g.	g.
Protein	28.50	15.28	10.60	8.84	8.86
Asparagine	(0.32)?	5.54	7.86	8.77	9.92
Amino acids	(2.52)?	7.63	10.19	10.90	10.57
Nitrogenous bases	2.25	3.52	2.62	1.55	1.50
Starch	37.82	17.44	9.93	3.94	2.59
Soluble carbohydrate	5.59	8.75	7.67	6.27	4.05
(glucose therein)	0	(2.43)	0	0	0
Ether extract	0.80	1.31	1.20	1.11	1.07
Ash	3.27	3.27	3.27	3.27	3.27
Crude fibre	6.64	7.70	9.15	9.65	10.98
Hemicellulose	4.70	5.25	5.80	6.10	6.40
Total loss of solids (respiration)	0	15.95	24.26	30.86	34.09
	92.41	91.67	92.55	91.26	93.30
Undetermined	7.59	8.33	7.45	8.74	6.70

the vetch seedlings was less intense than in those of the lupin, and Prianischnikow pointed out the analogy with the known protein-sparing action of carbohydrates and fats in the animal body. But in spite of the large reserve of various carbohydrates, protein breakdown in the early stages had been very rapid. He thought that this might occur because certain of the seed reserve proteins were more readily attacked by the enzymes than others; alternatively, it was possible that protein decomposition depended on the total amount of protein present, so that when this was large, as in the early stages, the decomposition would be more intense. Prianischnikow was suggesting here—in elementary terms—the "mass action" hypothesis recently formulated by Paech (200) to explain the regulation of protein decomposition and synthesis in plant cells.

Paech claims that his general conclusions are applicable to all parts of the plant at all stages of growth, and they are thus pertinent to the present discussion on seedling metabolism. They are of especial importance, however, in studies of the protein metabolism of leaves and, as most of his evidence is drawn from experiments with these organs, it will be more appropriate if I defer a full discussion of his hypothesis until Chapter XI. Nevertheless, he made two interesting experiments with seedlings, and as the conclusions to be drawn from them throw considerable light on the question both Schulze and Prianischnikow found difficulty in explaining, namely, the factors that control the breakdown of protein in the *early* stages of seedling growth, it will be convenient to give a short account of them here.

From an extensive review of the literature and from certain of his own experiments with leaves, Paech put forward the hypothesis that in plants the control of protein in the intact cell towards synthesis or decomposition is not determined by the nutritive ratio (Schulze) or by the amount of protein present (Prianischnikow), but takes place through the amounts of chemically active carbohydrates (monoses*) and active nitrogen (ammonium salts, amides and to a lesser extent amino acids and bases), so that an increase or decrease of the *component present in lesser amount* will bring about equivalent protein changes. Large amounts of monoses* and active nitrogen can thus never be present together in the same cell or within the same sphere of reaction.

His interpretation of protein decomposition in seedlings is as follows. On germination, the activation of the enzymes produces monoses and active nitrogen compounds from the reserve carbohydrates and proteins respectively. The growing parts exert the same "attractive force" on the soluble nitrogen compounds as they are

* As I show in Chapter XI, Paech's assumption that monoses may be a component of such a system is incorrect. The nitrogen-free products concerned—possibly α-ketonic acids—must be in nearer chemical relation with the amino acids than are the sugars. Such products, nevertheless, might, in some cases, be readily metabolised from glucose, and this would appear to be so in the two experiments about to be described.

known to do on water, mineral salts and soluble carbohydrates, so that both the monoses and active nitrogen compounds are drawn away from the reserve organs. In the case of wheat seedlings, the carbohydrates greatly predominate, so that the small amount of active nitrogenous compounds (the lesser component in terms of his hypothesis) is swept out of the endosperm with the larger amount of monoses, and this leads to continued protein decomposition. To illustrate this point, wheat grains, the proteolytic enzymes of which are known to become active within 24 hours of soaking, and to remain so during germination, were soaked in nitrogenous nutrient solutions for 24 hours and then grown for 4 days in the dark on blotting paper soaked in the same solution, controls being similarly treated with water. The embryos and remainder of the grains, without the testas, were analysed separately.

TABLE 22.
(From Paech, 200.)

Etiolated seedlings of *Triticum sativum*	After 5 days' growth			
	In 100 grains without embryo		In 100 embryos	
Nutrient solution	Protein-N	Soluble-N	Protein-N	Soluble-N
	mg.	mg.	mg.	mg.
H_2O	35.7	12.4	23.38	14.78
0.5% $NaNO_3$	47.3	16.7	25.82	27.44
H_2O	33.98	13.4	21.97	12.28
0.5% urea	45.66	31.66	23.91	41.66

The demand of the growing parts for active nitrogen being now in part satisfied from other sources, the call on the endosperm was reduced and the protein decomposition therein was checked. The protein level in the wheat endosperm therefore, which is rich in carbohydrate, really depends on the nitrogen requirements of the embryo, for Godlewski (87) showed that feeding with glucose brought about no change.

With low-carbohydrate seeds such as lupins, different conditions apply. Here the minimum component will be the monose derived from the small amount of insoluble carbohydrate reserves, and in this case it is an alternative source of active monose which should depress the protein decomposition in the cotyledons.

TABLE 23.

(From Paech, 200.)

Etiolated seedlings of *Lupinus albus*	After 14 days' growth			
	Cotyledons of 6 plants		Axial organs of 6 plants	
	Protein-N mg.	Soluble-N mg.	Protein-N mg.	Soluble-N mg.
N-free nutrient solution	19.82	23.72	12.82	81.54
N-free nutrient solution + 3% glucose	46.82	36.01	11.98	49.67

Seeds of the white lupin were germinated in water until the rootlets were about 1 cm. long, then transferred to nitrogen-free nutrient solutions, or nitrogen-free nutrient solutions with added glucose, and grown for 14 days in the dark. The results were as expected; the young plants which could obtain the major part of their carbon requirements through the roots demanded less from the cotyledons, and the extent of protein breakdown therein was much reduced. The enormous protein conversion in lupin seedlings therefore, which becomes very active even in the earliest stages, is due to the fact that the small amount of monose produced from the very limited carbohydrate reserve is rapidly translocated out of the cotyledons to the growing parts.

It would appear that Paech's views which, in effect, make the translocation stream responsible for the nitrogen metabolism between organs, are in harmony with the observations of earlier workers on protein decomposition in seedlings.

6. *Protein Regeneration in Seedlings.*

Schulze's later views on this question, which were based on his new concept of the secondary production of asparagine nitrogen and glutamine nitrogen from amino acids, brought him into much closer agreement with Pfeffer and Borodin than he had been at any time since their so-called hypothesis had been first formulated. These views were first set forth in 1898 (279, 280) and were again given, with but little change, in 1900 (281), 1903 (294), 1906 (285) and 1911 (286).

The amino acids and (possibly) the amides resulting from protein decomposition in the endosperm or cotyledons were transported, together with soluble carbohydrates, to the growing parts. There the amino acids underwent a secondary conversion to provide nitrogen for the elaboration of asparagine or glutamine, both of which substances were then used—in amount depending on the available carbohydrate (glucose) present—to regenerate protein. In conformity with his earlier views (cf. p. 23), he was still prepared to admit that the amino acids themselves, probably without the aid of glucose, might also take part in protein synthesis, but, for the many reasons set forth below, he did not now consider that this was a normal procedure.

In etiolated seedlings the growing parts contain a small and steadily diminishing amount of certain amino acids, but in proportions which are usually very different from those characteristic of the hydrolysis products of the seed reserve proteins (281). This was due to their secondary conversion, at rates characteristic of each particular amino acid, and of the plant species, to give asparagine (or glutamine). The same holds for seedlings that are grown in the light, the disappearance of the amino acids from the growing parts being, in the later stages (of *seedling* growth), even more marked (294). In this case, however, protein regeneration will eventually predominate, but even so there may be, at the same time, a simultaneous increase in the content of asparagine (cf. table 9, and pp. 22–24). It might be thought that the amino acids were being directly utilised here for protein synthesis, but in Schulze's opinion (contrary to his earlier view) the evidence for this was not clear, even though he had himself shown (282, 283) that leucine and tyrosine could serve as nitrogen sources for the growth of higher plants, and Treboux (324) had brought forward confirming evidence. If, on the one hand, asparagine be the chief and perhaps the only nitrogen containing substance (in the case of seedlings germinated without external supplies of nitrogen) used for protein regeneration—which was the view he now held—then some physiologically active nitrogen-free substance such as glucose would also be required for this purpose. That protein regeneration from these two products could actually take place, e.g. when etiolated seedlings were exposed to the light, had been abundantly verified by many experiments since the time of Pfeffer and Borodin (cf. table 8, stage 3). If, on the other hand, regeneration can be brought about directly from amino acids such as leucine or tyrosine, then other relationships must hold and nitrogen-free material such as glucose would either not be required, or, if so, then only in smaller amount. Yet all the available evidence showed that protein regeneration did not take place at all readily *except in the presence of soluble carbohydrates* and was most active in organs such as leaves which are rich in these substances. In his opinion, therefore, the amino acids underwent a secondary change to provide asparagine (or glutamine) prior to the regeneration of protein.

In support of this view he drew attention to his experiments with Castoro (294). In white lupins germinating in the light, there is an enormous conversion of protein nitrogen to asparagine nitrogen during the first 14 days' growth (cf. pp. 55, 56). This asparagine must be used later for the regeneration of protein in the developing plant (cf. table 27, in which the observed period was 28 days) and appropriate analyses showed that, while the leaflets were rich in protein but relatively poor in asparagine and other soluble nitrogenous products, the stems, which must supply the material used for protein synthesis in the leaflets, were extremely rich in asparagine. Many similar analyses quoted by Schulze and also by Prianischnikow confirm the high asparagine content of petioles and stems.

TABLE 24.
(From Schulze and Castoro, 294.)

14-day seedlings of *Lupinus albus* grown in light

	Cotyledons	Plumules	Stems	Roots
		In percentage of dry weight		
Total-N	8.83	6.57	6.77	5.40
Protein-N	2.44	4.11	1.56	1.87
Asparagine-N	3.73	1.41	4.48	2.17
Other soluble-N (diff.)	1.66	1.05	0.73	1.36

It was the undoubted ease with which all, or nearly all, of the amino acids arising from the breakdown of the reserve proteins can undergo a secondary change to provide nitrogen for asparagine (or glutamine) synthesis and the absence, or virtual absence, of many of these amino acids in those (new) parts of the plant wherein protein synthesis must take place (281, 294), which led Schulze in 1898 (280) to change the views he had previously held and to suggest that asparagine (or glutamine) was more suitable for protein synthesis than amino acids because *it was more easily available* for this purpose. He considered that his experiments on the ripening of seeds, to be discussed later, supported his new contention; in addition, he pointed out that the chemical constitution and properties of these two amides marked them out as material much more suitable for translocation of the nitrogen required for protein synthesis than the direct decomposition products, such as leucine and tyrosine, themselves (281).

Schulze thus came to regard asparagine (and glutamine) production as the first phase of protein regeneration and, in 1898 (279, 280), he put forward a new interpretation of one of his earlier experiments with yellow lupins (cf. pp. 22–24).

TABLE 25.

(From Schulze 268 and 279.)

Lupinus luteus	In percentages of total nitrogen	
	Seedlings which had been kept in the dark for 10 days	Seedlings kept in the dark for 10 days then exposed to light for 3 weeks
Protein-N	25.2	33.2
Asparagine-N	34.2	42.0
Other soluble-N (diff.)	40.6	24.8

He had previously held that the increase in protein nitrogen had taken place at the expense of the other soluble nitrogenous products (amino acids) present, but he now considered that this conclusion was not valid. It was asparagine itself which had been utilised in protein formation, and in this particular case the effect on the nitrogen partition had been masked by the fact that the first phase of protein regeneration (the conversion of amino acids to provide asparagine) had been more active than the second (the condensation of asparagine and glucose to give protein), due to the limited supplies of the latter substance available. He considered that the experiment supported his contention that asparagine production from amino acids was dependent on the state of development of the plant. Etiolated seedlings often contained more asparagine than those growing in the light, but *at the same stage* the ratio of asparagine nitrogen to other soluble (amino acid) nitrogen was always greater in the latter case. This was because the obligatory metabolic processes proceed more rapidly in plants growing under normal conditions, so that the conversion of amino acids to provide asparagine nitrogen (or glutamine nitrogen) for subsequent protein regeneration was encouraged.

In none of the above-mentioned summaries of his views does Schulze discuss the mechanism of protein regeneration from asparagine and glucose. There is no doubt, however, that he regarded asparagine and glutamine as vehicles for the nitrogen requirements of protein synthesis and as from 1888 onwards he had considered that the final stage in the secondary conversion of amino acids to asparagine might be a condensation of ammonia with succinic or malic acids, it is pertinent to recall his early statement of 1878 (cf. p. 23) that if asparagine could be used in protein regeneration it would first be broken down to one or other of these acids and ammonia, the latter being used subsequently in synthetic processes. Such views, if he held them in 1898, would antedate Prianischnikow's α and ω hypothesis (see p. 103) by many years; I doubt, however, if this were actually the case, for he did not again refer to his statement of 1878 in any of his later publications, and to those at all familiar with his work this can be taken as proof either that he had forgotten it, or, as is more likely, that in his maturer judgment he regarded the whole question as too highly speculative to warrant serious discussion.

While admitting that Schulze's insistence on the availability of asparagine (or glutamine) for synthetic purposes was justified by many observations, including his own, since the time of Pfeffer, and that it is in keeping with our modern views on the intermediary role of ammonia in protein metabolism, I cannot agree with the contention, which his later statements definitely imply, that amino acids, if available for the purpose, would not be used *directly* for protein regeneration as readily as these two amides themselves. I am inclined to think that the overriding importance which he attached to these latter substances in this connection can be traced to his inability to appreciate the full physiological significance of the secondary conversion of amino acids to ammonia. This change, which in his later years he agreed was an oxidation, was to him nothing more than a link in the chain of reactions by means of which protein in one part of the plant was transported to another.

$$\text{Protein} \xrightarrow{\text{enzymes}} \text{Amino acids} \xrightarrow{\text{oxidation}} \text{NH}_3 \xrightarrow{\text{organic acids}} \text{Asparagine} \xrightarrow{\text{glucose}} \text{Protein}$$

In other words he was unable to escape completely from the influence of the old Pfeffer hypothesis, which, in his early working years, had so completely dominated all views on protein metabolism in the plant. Asparagine production was the *first phase of protein regeneration,* and he (281) opposed the idea, first stated in definite terms by Prianischnikow in 1899 (225), that the secondary conversion of amino acids to ammonia might be due, in certain cases, to metabolic processes (e.g. respiration) quite unconnected with protein regeneration.

This new suggestion of Prianischnikow was formulated in 1895 soon after his return to Moscow from Schulze's laboratory. He recalled Boussingault's earlier view that asparagine production in the plant was analogous to urea production in the animal body (cf. p. 4), both being due to the "burning up" of protein; he had also noticed an early paper by Laskowsky (132) showing that a rise in the temperature of germination, which was known to increase the rate of respiration, led to increased asparagine accumulation in the seedlings. Suspecting that there might be some intimate connection between asparagine formation and respiration, Prianischnikow, during the next four years, carried out a comprehensive series of experiments to throw more light on the subject.

Seeds of the pea (*Pisum sativum*), broad bean (*Vicia Faba*) and the yellow lupin (*Lupinus luteus*) were germinated in the dark and samples taken at frequent intervals for analysis (225). The daily *rates* of protein decomposition and of asparagine accumulation were then calculated, and it was found that both reached a maximum at about the same time (7–9 days). At a later stage, the rate of protein breakdown fell much more steeply than did that of asparagine accumulation, and from this observation Prianischnikow inferred that asparagine was then being produced, in part, from other soluble nitrogenous products present, namely, the amino acids which had been derived from the primary breakdown of the reserve proteins. Figure 1, taken from one of his later summaries (230), illustrates these points.

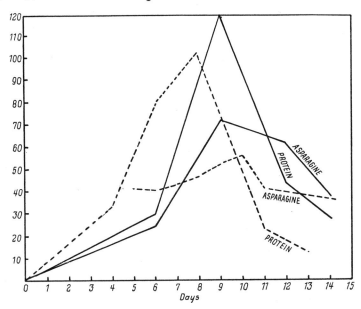

Fig. 1. *From Prianischnikow, 230.*

He had also measured the daily output of carbon dioxide. This remained high throughout, and actually reached its maximum some days later than the other two factors mentioned above. Since the respiratory quotient varied somewhat throughout the experiment, he could not be certain (in 1899) that the conversion of the amino acids to asparagine (really the transfer of nitrogen from the former to the latter, as he was careful to point out) was necessarily an oxidation, but the observation strengthened his surmise that the asparagine production might have been the result of metabolic processes which had no direct connection at all with protein regeneration. That these processes were connected with respiration rather than with growth was the inference he drew from his repetition (227) of Laskowsky's experiments. Samples of 8-day-old seedlings were placed in incubators kept at 20°, 28° and 34–36°, respectively, and grown for a further 5 days in the dark. The most favourable temperature for growth was 28°, whereas the energy of respiration and the rates of both protein decomposition and asparagine production increased with rise of temperature to 36°.

TABLE 26.

(From Prianischnikow, 227.)

	In percentages of total seedling-N		
	At 20°	At 28°	At 34–36°
Protein-N decomposed, per diem	1.42	3.30	4.22
Asparagine-N formed, " "	0.57	1.20	1.48

Seedlings 4 and 9 days old gave similar results, and Prianischnikow concluded that asparagine production from amino acids was due to living processes in the plant *connected with respiration rather than with growth*. Asparagine (and glutamine) therefore would tend to accumulate in the regions of most intensive metabolic activity, i.e. the growing parts.

In his next experiment (226), seedlings of the yellow lupin, runner bean and pea, among others, were grown in the light for longer periods than Schulze had usually employed.

In agreement with Schulze's findings, the extent of protein breakdown and of asparagine production was governed in each case by the nutritive ratio. Protein regeneration became active on leaf development between the tenth and fifteenth days and, as in all cases except that of the lupin, the major loss was shown by "other soluble-N." Prianischnikow, in keeping with Schulze's original suggestion of 1878 (cf. p. 23), inferred that the regeneration had taken place *preferentially at the expense of the amino acids*. He queried Schulze's later contention that asparagine and a physiologically active carbohydrate such as glucose were the more suitable raw materials for

TABLE 27.

(From Prianischnikow, 226.)

Seedlings growing in the light	Nitrogen figures in percentages of total seedling-N			
Lupinus luteus				
Age in days	7	14	21	28
Protein-N	37.67	52.01	68.77	70.00
Asparagine-N	30.35	21.86	13.56	12.92
Other soluble-N (diff.)	31.98	26.13	17.67	17.08
Phaseolus multiflorus				
Age in days	0	7	14	21
Protein-N	82.68	52.6	68.22	73.60
Asparagine-N	3.65	11.2	7.77	8.52
Other soluble-N	14.77	36.2	26.01	17.88
Pisum sativum					
Age in days	0	10	20	30	40
Protein-N	81.86	52.37	80.53	71.21	73.54
Asparagine-N	5.58	13.50	12.44	10.98	11.88
Other soluble-N (diff.)	12.56	34.13	7.03	17.81	14.58

this purpose; Seliwanow (306) had confirmed Schulze's own original observation (p. 34) that both these products were present in potato tubers and many similar occurrences could be cited. He also called attention to the fact that no case was known in which the feeding of glucose to etiolated seedlings had favoured protein regeneration. (Paech's observation, table 23, p. 70, is in keeping with this contention, and according to his hypothesis it is attributable to a reduced flow of active nitrogenous compounds from the seed reserves.)

Finally, in confirmation of his view that protein regeneration could take place from amino acids, *even to the exclusion of any asparagine present,* he repeated and confirmed Zaleski's (367) observation that, if bulbs of the onion (*Allium cepa*) were allowed to sprout over distilled water in the dark, an increase in protein nitrogen took place, preferentially at the expense of the amino acids. The following refers to a similar but later experiment of Zaleski and Shatkin (370), in which the amino nitrogen was determined by Van Slyke's method.

TABLE 28.

(From Zaleski and Shatkin, 370.)

Bulbs of *Allium cepa*	In percentages of total-N	
	Control	Grown for 25 days
Protein-N	44.4	63.4
Peptone-N	11.0	9.0
Ammonia-N	4.5	3.0
Amide-N	11.4	10.7
Basic-N	3.5	2.7
Monoamino-N	26.0	10.0

Incidentally, it may be mentioned that the nitrogen (and other) changes taking place during the forcing of these bulbs have been recently investigated in much greater detail by Rahn (237).* At the beginning of growth the inner scales lose protein more rapidly than the outer scales; the soluble nitrogen also falls rapidly, so that all the hydrolysis products are taken up by the young leaves. Examination of the soluble nitrogen shows that the α-amino nitrogen (Van Slyke) disappears completely from the inner scales in two weeks and, during the same period, falls to two-fifths in the outer scales. On the other hand, the amide nitrogen falls slowly in both cases (fig. 2). Rahn concluded that, as in seed germination, the α-amino nitrogen is the important factor in translocation to the growing parts (leaves), where it is used for protein regeneration.

Prianischnikow's final conclusion in 1899 was that as-

* The various *titration* methods used by Rahn to fractionate the water-soluble nitrogenous products have been recently criticised by Richardson (241).

paragine and glutamine, in so far as they arose from the secondary conversion of amino acids, were to be regarded as nothing more than by-products of metabolism which serve as a storage of nitrogen, although, under favourable conditions, they could contribute, as readily as the amino

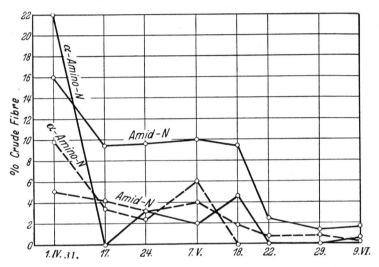

Fig. 2. From Rahn, 237. —— *inner scales* --- *outer scales*.

acids themselves, towards protein regeneration. The accumulation of asparagine in leaf petioles, therefore, was due to the fact that the substance was not immediately required for this purpose. This view he maintained in the face of Schulze's criticisms (281) and interpreted the latter's own lupin data (table 9, p. 24) accordingly. In his opinion, the protein regeneration in these seedlings had taken place chiefly at the expense of the amino acids—which were still being produced by decomposition of the cotyledon reserve proteins—and the concomitant increase in asparagine took place since the metabolic processes involving the production of ammonia from amino acids were still very active, in spite of the onset of photosynthesis.

I shall discuss in Chapter IV the able manner in which

Prianischnikow was later to develop the concept that plants avoid the toxic action of ammonia—whether produced from the metabolism of amino acids or whether supplied from external sources—by elaborating asparagine (or glutamine), and that ammonia is the α and ω of protein metabolism, for, via amino acids, it is the necessary precursor of protein synthesis as well as the end-product of protein degradation (cf. pp. 98 *et seq.*).

In retrospect, the conflict between Prianischnikow and Schulze on the question of protein regeneration in seedlings appears to have been nothing more than a difference of opinion as to the "availability" of various products for this purpose, and the undoubted bias that Schulze had in favour of asparagine can be traced to his frequent choice for experimental purposes of lupins, in which the protein conversion to give this amide is phenomenally large. In judging them, we must remember that at the time (1900) not only were current ideas of protein structure, and therefore of possible mechanisms of protein synthesis, still rudimentary (the peptide hypothesis was enunciated in 1903), but the amino acid composition of the proteins of the growing parts—especially of the leaflets—was as yet unknown, so that neither investigator was in a position to consider the question of protein regeneration in relation to *amino acid requirements*. During recent years, however, my colleagues have provided us with a number of leaf protein analyses (tables 49 and 53, Chapters VI and VII) and within certain specified limits it is now possible to review the earlier data of Prianischnikow and Schulze from such a standpoint.

For a valid correlation of the changes involved it is necessary to know at each stage the amount of protein that remains in the cotyledons, and the amount synthesised in the new growth (which will include an increasing proportion of leaves). Unfortunately, Prianischnikow did not make separate analyses of the cotyledons and new growing parts of his seedlings, and I can only assume, since external sources of nitrogen were excluded, that the major part of the reserve protein had been decomposed within the first 10–15 days. I shall also assume that protein synthesis takes place by repeated condensation of amino acids to form peptide chains, and that the required amino acids are either already available as decomposition products of the seed reserve proteins or can be readily synthesised, according to the general schema (reaction 6) discussed in Chapter V, from α-ketonic acids and "available" ammonia, the latter being provided by an excess of asparagine, glutamine or any amino acid present.

Let us consider first of all Prianischnikow's experiments with pea seedlings, which he regarded as noteworthy in that protein regeneration (somewhat irregu-

TABLE 29.

Amino acid analyses of certain seed and leaf proteins.

(Figures given are in percentages of protein.)

	Seed proteins			Leaf proteins		
	Pisum sativum	Phaseolus multiflorus	Lupinus luteus	Medicago sativa	Phaseolus multiflorus	Trifolium repens
	(1)	(2)	(3)	(4)	(5)	(6)
Ammonia	1.9	1.69	2.75	1.07	1.11	1.17
Aspartic acid	5.1	5.2	9.4	7.6	8.4	7.6
Glutamic acid	18.1	14.5	23.8	11.8	12.0	11.8
Arginine	9.5	4.9	11.0	8.0	7.9	8.1
Histidine	2.0	2.6	2.0	1.5	1.6	1.8
Lysine	4.8	4.6	2.3	6.2	5.4	5.8

(1) According to Osborne and Campbell (191) the three constituent proteins legumelin, legumin and vicilin are present in the ratio 2:5:5. The analysis quoted is a composite one from those of Osborne and Clapp (192) and Osborne and Heyl (195).

(2) Phaseolin. It represents only a fraction of the total seed protein (Osborne and Clapp, 193).

(3) Conglutin represents 80% of the total seed protein (Osborne and Campbell, 190). New analysis recently obtained in this laboratory by Dr. Tristram.

(4) and (6) from table 53; (5) from table 49.

lar, due to uneven leaf development) took place almost exclusively at the expense of the amino acids. Here the seeds contain a fairly large reserve of carbohydrate, so that the obligatory metabolic processes connected with e.g. respiration will not call for a strong conversion of amino acids to ammonia and hence will not lead to much asparagine production. If we compare the amino acid composition of the total protein present in the cotyledons with that of the protein present in a typical leguminous leaf—and the major part of the protein regeneration must take place in the latter organs—we see at once that the breakdown of the former protein will provide bases and aspartic acid (or asparagine) in proportions very close to those required for *direct participation* in protein regeneration. Glutamic acid (or glutamine) will be produced in greater amount than required for this purpose and the excess may very well contribute towards asparagine formation (cf. p. 95). Within the limits set by the few amino acids that can be legitimately considered in this connection therefore, it is apparent that protein regeneration can take place in large part without calling in the aid of "available ammonia," so that the asparagine produced by metabolism of certain of the primary decomposition products should remain more or less unchanged in amount. This is in agreement with Prianischnikow's experimental data.

According to Osborne, the seeds of the broad bean (*Vicia Faba*) contain three proteins similar to those of the pea; Prianischnikow's observation therefore that in these seedlings very little protein regeneration was apparent after 40 days' growth, although there was a small *increase* in asparagine, can be interpreted in a similar way. Protein breakdown in the cotyledons was balanced by protein regeneration in the leaves and the amino acids provided by the former operation were in large part utilised *directly* in the latter.

In the case of the runner bean (*Phaseolus multiflorus*), the conditions were slightly different. The excess glutamic acid might, as before, be metabolised to asparagine, but in addition 3 units of arginine (12 units of arginine nitrogen) must be provided for the protein regeneration. There will be a definite call therefore on the "available ammonia" and the content of asparagine should slowly fall. Again this is in keeping with Prianischnikow's experimental data.

Lastly, we have the case of Prianischnikow's and Schulze's lupin seedlings. These seeds contain very little carbohydrate reserve, so that the obligatory metabolic processes will bring about a large conversion of amino acids to asparagine. For this purpose there is an excess of arginine (rich in nitrogen) and a larger excess of glutamic acid. During regeneration 3 units of lysine (6 units of lysine nitrogen) must be synthesised and there will be again a steady call on the "available ammonia." Thus, within the limits that I have already mentioned, it seems to me that Schulze's views can be brought into harmony with those of Prianischnikow—which are undoubtedly correct—if we consider protein regeneration *not only in terms of amino acid availability but also in terms of amino acid requirements.*

In the foregoing discussion on protein regeneration, I have given full weight to the views of Schulze and Prianischnikow, for their conclusions were based, in large part, on evidence derived from actual experiments *with seedlings*. I have not attempted to introduce modern views on the mechanisms of biological synthesis, degradation and interconversion of amino acids, based—and I should like to emphasise the point—to a large extent on experiments with animal tissues, for however straightforward an explanation these might, at first sight, appear to give of the role of asparagine, glutamine and amino acids in plant protein metabolism, the issue is, in reality, complicated by the presence of reactive carbohydrates and organic acids (especially), as will appear from the subject matter of Chapters V, IX and X. It will be more convenient, therefore, to defer further discussion on protein metabolism in seedlings until Chapter V, and I close the present one with a summary of what can be regarded as a modern interpretation of the views of Schulze and Prianischnikow.

(1) On germination, the reserve proteins undergo enzymic digestion to give, as primary products of hydrolysis, asparagine, glutamine and the usual series of amino acids, including aspartic acid and glutamic acid if these be combined as such (i.e. in the unamidised form) in the intact protein molecule. The reserve carbohydrates also undergo enzymic breakdown to give simpler products, including monoses.

(2) The decomposition of the reserve proteins is controlled by the outflow from the cotyledons or endosperm, as the case may be, of either monoses ("active" carbohydrates) or amino acids, etc. ("active" ammonia), depending on which component is present in the lesser amount (Paech).

(3) The primary products of reserve protein decomposition are transported to the growing parts, where they become available for protein regeneration.

(4) Obligatory metabolic processes, which Prianischnikow connected with respiration rather than with growth, bring about a secondary decomposition of certain of the amino acids, which results in the formation of asparagine, glutamine or both these amides.

(5) Protein synthesis in the growing parts takes place at the expense of the amino acids translocated from the reserve organs, and of the acid amides resulting from secondary changes.

(6) The proteins of the growing parts may have a different amino acid composition from those of the reserve organs. Protein regeneration, therefore, in the absence of external sources of nitrogen, may necessitate an interconversion of amino acids (really a transfer of nitrogen from one nitrogen-free acid to another), and this probably takes place through the agency of asparagine and glutamine.

CHAPTER IV

ASPARAGINE AND GLUTAMINE FORMATION IN SEEDLINGS

1. *Introduction.*

IN many researches dealing with amide or protein metabolism in the plant, it has been tacitly or perhaps unwittingly assumed that, if protein nitrogen disappears and is replaced by amide nitrogen, then protein has been metabolised to asparagine. This is an instance of loose thinking, for a little consideration will show that the assumption is true only of those (rare) cases in which it can be shown that the amides are primary products of protein decomposition. If the amide nitrogen be of secondary origin, as is usual, then the assumption is, in part, a false one, for a demonstration that protein nitrogen has been transformed to amide nitrogen provides no evidence whatever as to the origin of the remainder of the amide molecule. The error, which is prevalent in the literature, can be traced to the universal custom of evaluating proteins and amides in terms of nitrogen, so that by degrees the mind becomes neglectful of the fact that both these products contain another element—carbon—which is of equal importance in plant metabolism as is the nitrogen itself. The nitrogen partition, which is such a convenient form of presentation when discussing metabolic changes, helps to obscure the issue, and even Schulze, who was generally so very meticulous in drawing conclusions from his experimental data, fell into the snare when challenged by Müller as to the origin of the asparagine in his lupin seedlings.

This point is one of great significance in certain studies which I shall be discussing later, for if, as we now believe, protein breakdown leads ultimately to ammonia and a series of α-ketonic acids, the carbon skeletons of the two respective amides may indeed arise through further metabolism of these ketonic acids, but may equally well be derived from the metabolism of non-nitrogenous plant materials such as carbohydrates or fats. Recent work has indeed tended to emphasise this latter alternative, and if it be true, we are left with the difficult task of deciding the ultimate fate of the ketonic acids. I propose therefore to discuss both alternatives in some detail.

2. *The Oxidation of Amino Acids to Ammonia.*

We saw in Chapter II (p. 48) that while Schulze, as early as 1888, had put forward the suggestion that the amino acids resulting from primary breakdown of protein in seedlings might undergo a secondary decomposition to provide nitrogen-rich residues, or even ammonia, for asparagine production, and in 1892 had discussed the possibility that, in certain instances, this secondary decomposition might be an oxidation, he would not accept the view that the amino acids normally underwent oxidation to give ammonia until experimental evidence in support of it had been put forward soon after 1900. Prianischnikow (225), however, as we have already seen (p. 76) was more willing to speculate at this period than Schulze, and, since asparagine contained two atoms of nitrogen, he inferred that its synthesis required, in the first place, one molecule of aspartic acid, which might well be a product of primary protein decomposition, and, in the second place, one molecule of ammonia, which might arise through some deep-seated change—probably an oxidation—involving another amino acid; the ammonium aspartate thus produced undergoing dehydration to give asparagine. This view, of course, recalls the earlier (vitalistic) speculations of Löw (p. 51).

Prianischnikow pointed out that such a synthesis was analogous to the production of urea in the animal body; in both cases the protein first underwent digestion (hydrolysis) to give amino acids, which were then conveyed to other organs (in the case of the seed, to the axial organs) wherein oxidation to ammonia and subsequent synthesis, to urea or asparagine, respectively, took place.

The main weakness of Prianischnikow's hypothesis was his assumption that the aspartic acid required for the asparagine synthesis could arise through primary protein breakdown; Schulze had always doubted whether, except in small amount, this were possible, and we now know that his doubts were justified. Be that as it may, the hypothesis was undoubtedly of great value, for it introduced for the first time *without any equivocation* the concept that asparagine production required the oxidation of amino acids to give ammonia, and it undoubtedly influenced researches conducted in later years which definitely proved it. Prianischnikow, at the time, admitted the difficulty of effecting such an oxidation *in vitro*, and was forced to rely on the evidence of Palladin (which we have already seen was disputed by Clausen) that anaerobiosis depressed asparagine formation (p. 51). Five years later, however, in 1904, he had no further doubts (228), for he was able to quote an experiment of Demjanow showing that the oxidation of leucine with permanganate gave valeric acid, carbon dioxide and ammonia.

$$\begin{array}{c}CH_3\\ \diagdown\\ CH.CH_2.CH(NH_2).CO_2H + O_2 =\\ \diagup\\ CH_3\end{array} \begin{array}{c}CH_3\\ \diagdown\\ CH.CH_2.CO_2H + CO_2 + NH_3\\ \diagup\\ CH_3\end{array}$$

He was also to refer to the work of Butkewitsch (27), published two years previously, showing that, in seedlings anaesthetised with toluene vapour, synthetic processes ceased but protein decomposition went on uninterruptedly, carbon dioxide was released and ammonia accumulated instead of asparagine.

This latter evidence convinced Schulze that the breakdown of amino acids could give ammonia, but he was still unconvinced that the process was necessarily an oxidation. He agreed (285) that guanidine, which he had found some years before in seedlings of the vetch (275), might be an oxidation product of arginine, but many unknown bases were present in most of the seedling extracts he had examined, and he thought it possible that some of these might arise by simple decarboxylation, as they clearly did when the amino acids were acted on by bacteria. Accord-

TABLE 30.

(From Butkewitsch 26.)

Lupinus luteus	Percentages of total-N of seedling	
	Untreated	3 days in toluene vapour
Soluble-N	32.1	34.5
Amide-N	6.1	2.6
Ammonia-N	0.9	8.3

ingly, he examined (284) several seedling saps with the object of isolating tetramethylenediamine (from arginine *via* ornithine), pentamethylenediamine (from lysine) and phenylethylamine (from phenylalanine), with, however, negative results. Further (biological) evidence in favour of the oxidation hypothesis was brought forward about this time by Suzuki (321), Butkewitsch (26) and Godlewski (88), but Paech (200) has recently shown, in agreement with the earlier view of Clausen (p. 51), that under conditions of anaerobiosis, plants rapidly die, so that most of the changes observed by these workers must have been *post mortem*.

I think, however, that we can now regard such evidence as immaterial, since recent work (Krebs, 123, 124) has shown that the first stage in the biological decomposition of amino acids is an oxidation to the corresponding α-ketonic acid with the production of ammonia. The evidence for this is fully discussed later, in Chapter V.

$$R.CH(NH_2).CO_2H + O = R.CO.CO_2H + NH_3$$

The ammonia resulting from the amino acid oxidation, however, is not always utilised in the synthesis of amides; Ruhland and Wetzel (255, 256, 257), for example, hold that in certain plants—those with a very acid sap—the normal metabolism, and also that under special experimental conditions, is associated with the formation of ammonium salts of organic, e.g. malic and oxalic, acids. They accordingly group plants under two headings: "amide plants" and "acid or ammonia plants," examples of the latter being *Begonia semperflorens* (the leaves of which are very acid, pH 1.53, and contain much ammonia, especially if cultured in the dark), and the petiole of rhubarb. The distinction was further emphasised by the work of Kultzscher (129), but has recently been queried by Schwab (304), whose more accurate analyses show that the "acid or ammonia plants" also produce considerable amounts of amides. According to Kultzscher, the leaves of *Oxalis Deppei* have a sap at about pH 1.5 and of *Pelargonium* at about pH 2.82.

TABLE 31.
(From Schwab, 304.)

	Number of days in dark	In percentages of total leaf-N			
		Protein-N	Ammonia-N	Glutamine-N	Asparagine-N
Leaves of *Oxalis Deppei*	0	83.0	4.46	1.2	2.08
" " " "	2	59.6	9.23	10.1	6.03
" " " "	5	28.2	31.3	17.9	2.0
" " *Pelargonium peltatum*	0	80.8	4.00	0.1	0.94
" " " "	2	68.8	4.83	3.79	2.78
" " " "	4	50.0	20.1	0.93	10.78
" " " "	5	37.4	30.9	0	9.05

3. *The Possible Secondary Production of the Whole Asparagine Molecule from Protein.*

Schulze (p. 48), when attempting to prove that the asparagine which accumulated in his lupin seedlings (grown under conditions which precluded the entry of extraneous nitrogen) came from protein, begged the question by considering only the transfer of nitrogen. To the best of my knowledge, the origin of the carbon skeleton of the asparagine produced during the etiolated growth of seedlings has never been directly investigated, so that no *ad hoc* evidence is available; but two of Schulze's papers have provided me with sufficient data for profitable discussion. These are set out in table 32. Certain assumptions have been made in calculating the weights of the different fractions, but the possible errors introduced will not affect the point at issue.

The non-nitrogenous reserves of the seed available for metabolism during germination are the hemicelluloses, which he showed were readily digested, water-soluble polysaccharides, and the fraction designated "other solu-

TABLE 32.

(Computed from Schulze, 279, 277.)

Lupinus luteus

Stage	Seeds (1)	Seedlings kept 15 days in dark (2)	Seedlings kept 15 days in dark and then 10 days in light (3)
	g.	g.	g.
Total dry weight	100	77.2	82.2
Protein (N × 6.25)	52.3	10.3	10.7
Nitrogenous bases (N × 5)	2.3	6.1	5.1
Asparagine	–	19.0	24.1
Other soluble nitrogenous compounds, assumed to be monoamino acids (N × 7)	1.1	16.7	9.8
Lipoid (ether extract)	6.2	4.0*	4.0*
Available N-free reserves { Water soluble polysaccharides 8.4; Other soluble N-free material 15.3; Hemicelluloses 9.6 }	33.3	16.3	23.7
Crude fibre	4.8	4.8	4.8

* Schulze does not quote values for seedling "ether extract" and those given are based on experiments in my laboratory with seedlings of *Phaseolus multiflorus* (113).

ble N-free material." This last may contain sugars and will include the small amounts of malic and citric acids Schulze found in both the seed and seedlings.

Let us consider some of the major changes that have occurred during growth in the dark for the first 15 days. In the first place, 22.8 g. of total solids have been lost through respiration or perhaps, in small part, by the formation of water in metabolic processes. In the second place, 42 g. of protein have disappeared, and of this amount about 22 g. are represented by the increases in "nitrogenous bases" and "other soluble nitrogenous compounds." These can be considered in large part as primary products of the protein breakdown and the origin of their carbon skeleton is therefore not in question. The remaining 20 g. have been metabolised beyond the amino acid stage, the nitrogen being incorporated in the newly formed asparagine, and the α-ketonic acids, in all probability, further metabolised to what I refer to as "N-free fragments." These latter, of course, are available for respiratory or synthetic processes. We are now in a position to draw up a simple balance sheet.

Balance sheet at stage 2.

To be accounted for:	Available for metabolism:
22.8 g. of total solids	2 g. lipoid
About 15 g. of N-free asparagine	About 17 g. of "N-free fragments" derived from protein
	17 g. of "available N-free reserves"

In spite of liberties which have admittedly been taken in computing the numerical values given above, two alternative conclusions stand out quite clearly. (1) If the "available N-free reserves" and lipoid have been preferentially utilised for respiratory processes, as some would hold, then the "N-free fragments" derived from the protein must have provided the major part of the asparagine carbon skeleton. (2) If the "available N-free reserves" have furnished the latter—and under different conditions, as I shall show later, we know that glucose can do this—

then the "N-free fragments" have been used in respiration. Schulze's data therefore provide an interesting confirmation for the view of Prianischnikow (p. 81) and the earlier suggestion of Boussingault (p. 4) that *asparagine production in seedlings is connected with respiration.*

Equally interesting are the deductions from the changes which occurred when the 15-day etiolated seedlings were grown for a further period of 10 days in the light. In this case, the fixation of carbon by photosynthesis introduces an unknown factor, but that need not prevent us from drawing up a balance sheet showing the changes that have occurred during this second period of growth.

Balance sheet of changes occurring between stages 2 and 3.

To be accounted for:	Available for metabolism:
About 5 g. of N-free asparagine	5.0 g. of total solids
7.4 g. of "N-free reserves"	About 8.5 g. of N-free fragments from "nitrogenous bases" and "other soluble nitrogenous compounds"

During a period of active photosynthesis (shown by the increase in total solids), it seems reasonable to assume that newly formed sugars have contributed in large part to the increase in "N-free reserves," in which case we cannot, as before, escape the conclusion that either the "N-free fragments" derived from the soluble nitrogenous products have been used in respiration, or they have become incorporated in the newly formed asparagine. The data of Schulze, therefore, support the assumption that in seedlings growing under conditions of carbohydrate deficiency, proteins can be used (indirectly) in respiration. Definite proof that this can take place in tobacco leaves, cultured in water for some days in the dark, is provided in Chapter X. Schulze's seedling data also show that the formation by secondary reactions of the whole asparagine molecule from protein is possible, and it thus becomes pertinent to enquire which of the many amino acids produced on protein digestion can readily function in this way.

AMIDE FORMATION IN SEEDLINGS

We have seen that the first stage in the metabolism of amino acids is probably an oxidative deamination with the production of the corresponding α-ketonic acid (p. 90). The next stage—decarboxylation—is generally considered to be brought about by carboxylase, the presence of which has been demonstrated in a number of higher plants (Zaleski and Marx, 369), the aldehyde produced being further oxidised to the corresponding acid.

$$R.CO.CO_2H \rightarrow R.CHO \rightarrow R.CO_2H$$

The amino acids will therefore be oxidised to non-nitrogenous acids containing one carbon atom less than their respective precursors. Glutamic acid will thus give succinic acid, which could be metabolised *via* fumaric acid or oxalacetic acid to asparagine, according to the scheme discussed in Chapter IX (reactions 8 and 9, pp. 190, 191).

It is interesting to observe that Damodaran and Nair (61) have shown that seedlings of three legumes contain an aerobic dehydrogenase which will oxidise $l(+)$-glutamic acid to α-ketoglutaric acid. Since this family of plants generally elaborates asparagine in preference to glutamine it is to be expected that the glutamine or glutamic acid produced on primary protein decomposition could be further metabolised to help provide the former amide.

The other non-nitrogenous acids no doubt undergo further oxidation with the production of shorter fragments, but three of them, arginine, proline and histidine might furnish succinic acid.

Arginine

$$HN:C(NH_2).NH.(CH_2)_3CH(NH_2).CO_2H \rightarrow HN:C(NH_2).NH.(CH_2)_3 CO_2H \rightarrow$$
$$HN:C(NH_2).NH_2 + HO_2C.(CH_2)_2.CO_2H$$
guanidine

This oxidation has been accomplished *in vitro* with barium permanganate by Kutscher (131), and led Schulze

Proline

Weil-Malherbe and Krebs (359) have shown that oxidative breakdown may occur in the kidney, with the production of glutamic and α-ketoglutaric acids. This amino acid might therefore act as a precursor of both glutamine and asparagine.

$$\begin{array}{c} H_2C-CH_2 \\ |\quad\quad| \\ H_2C\quad CH.CO_2H \\ \diagdown\diagup \\ N \\ H \end{array} \xrightarrow{O_2} \begin{array}{c} CO_2H.CH_2CH_2.CH(NH_2).CO_2H \\ \downarrow \\ CO_2H.CH_2.CH_2.CO.CO_2H \end{array}$$

Histidine

Edlbacher (71, 72) has concluded that the path of physiological disintegration of histidine proceeds, under the initial action of histidase, as follows:

and Castoro (294) to suggest that it might explain the presence of both guanidine and succinic acid in plants.

AMIDE FORMATION IN SEEDLINGS

The glutamic and succinic acids thus formed could act as precursors of the amides.

Reverting to Schulze's data for yellow lupins (pp. 92, 93), it is now possible to give some idea of the *amount* of secondary asparagine that could be formed in this way. A recent analysis of the lupin protein conglutin in my laboratory gave the following values:

	Per cent of protein
Glutamic acid	23.8
Aspartic acid	9.4
Arginine	11.0
Histidine	2.0
Proline*	2.6

The aspartic acid could be regarded as an immediate precursor of asparagine, even if it did not exist entirely in this form in the intact protein molecule, while calculation shows that, on a basis of 40 per cent protein decomposition, the other four amino acids could give 11 per cent of succinic acid. Such an amount, together with the preformed aspartic acid, could readily furnish the necessary 19 per cent of asparagine. It need hardly be emphasised that the above-mentioned hypotheses are purely speculative, especially as short fragments from the further oxidation of practically all of the amino acids might pass, *via* acetaldehyde to acetic acid, and hence by the Thunberg reaction to succinic acid.

$$\begin{array}{c} CH_3CO_2H \\ CH_3.CO_2H \end{array} \rightarrow \begin{array}{c} CH_2.CO_2H \\ | \\ CH_2.CO_2H \end{array}$$

They do, however, illustrate the possibilities of a direct synthesis of asparagine from protein, and emphasise that we have here a very important problem in plant biochemistry on which more experimental evidence is urgently required.

* Abderhalden and Herrick (1).

4. *The Accumulation of Asparagine and Glutamine in Plants Artificially Enriched with Ammonia.**

If the carbon skeleton of asparagine, and perhaps of glutamine, can be provided by certain nitrogen-free reserves, then plants which contain the necessary precursors should synthesise asparagine or glutamine when artificially enriched with ammonia. The truth of this was first demonstrated by Suzuki (319), who found that plants removed from the soil and placed in culture solutions took up ammonia and converted it to asparagine, especially if glucose were present in the culture solution. In no case was there any large accumulation of ammonia itself. These observations have been confirmed by Prianischnikow, who, from the results of a series of beautifully planned experiments, has been able greatly to extend our knowledge of the physiological role of ammonia and asparagine in plant metabolism.

Prianischnikow and Schulow (231) in 1910 repeated Suzuki's experiments under more easily controlled conditions. Samples of barley were germinated for 14 days in water and in 0.1 per cent ammonium chloride, respectively. Analyses of the seedlings showed that the ammonia taken up had been metabolised to asparagine.

Preliminary experiments with peas failed, but, when calcium carbonate or sulphate was added to the ammonium chloride culture solution, synthesis of asparagine was again demonstrated.

It will be seen that, in both sets of experiments, the increase in total nitrogen is closely paralleled by the increase in asparagine nitrogen, showing quite clearly that the ammonia from the culture solution had provided both the amino and amide groups in the asparagine molecule. As Prianischnikow emphasised, he had, for the first time, induced seedlings to build up asparagine under conditions clearly dissociated from a direct relationship to protein

* In the following account I have drawn largely on the excellent summary of Prianischnikow's work by Vickery *et al.* (336, p. 764).

metabolism, the amide having arisen *de novo* from ammonia and a non-nitrogenous precursor such as (for example) malic acid.

TABLE 33.

(From Prianischnikow and Schulow, 231.)

Seedlings of *Hordeum sativum* (vulgare)

Culture solution	H_2O	NH_4Cl
	Per 100 seedlings	
	mg.	mg.
Total-N	145.8	161.5
Protein-N	61.8	61.5
Asparagine-N	36.8	56.4
Ammonia-N	0.55	0.89

TABLE 34.

(From Prianischnikow and Schulow, 231.)

Seedlings of *Pisum sativum*

Culture solution	NH_4Cl	$NH_4Cl +$ $CaCO_3$	$NH_4Cl +$ $CaSO_4$
N taken up by seedlings	44.8	93.9	195.7
Increase in asparagine-N	25.0	116	182

Precisely similar results have recently been obtained by Vickery, Pucher and Clark (334) with respect to glutamine in beets from plots very heavily dressed with ammonium sulphate.

TABLE 35.

(From Vickery, Pucher and Clark, 334.)

Roots of *Beta maritima*	Control	Per kg. of fresh tissue	
		3 days after dressing of plot	9 days after dressing of plot
	g.	g.	g.
Soluble-N	1.96	2.10	2.16
Glutamine-N	0.37	0.54	0.58
Increase in soluble-N		0.15	0.21
Increase in glutamine-N		0.16	0.21

The effect is striking, for the whole of the nitrogen taken up has been metabolised to glutamine, showing that the non-nitrogenous precursor of this amide may also become available under conditions dissociated from protein metabolism.

These results are of the greatest importance, for they show *inter alia* that the second alternative I deduced from Schulze's lupin analyses (p. 73), i.e. that, on metabolism of the primary amino acids, the ammonia set free might condense with nitrogen-free reserve material to give asparagine, and that the residual carbon chains might be further oxidised and used in respiration, is very possible, although not necessarily so in that particular case.

Prianischnikow's interest in 1910, however, was directed towards the metabolism of ammonia. In 1895 (222), and again in 1899 (225), he had enlarged on Boussingault's original suggestion that asparagine formation in plants was analogous to urea formation in the animal body, and his new *direct* production of asparagine from ammonia led him to the further conclusion that both plants and animals avoid the harmful effects of ammonia in the cells by a dehydration process, namely, the synthesis of urea on the one hand or of asparagine on the other. Not until 1922 was he in a position to generalise more widely, for in some of his earlier experiments with Schulow they had been puzzled by the fact that certain plants accumulated small amounts of ammonia and did not metabolise it to asparagine (229).

Appropriate tests showed that the seedlings which did metabolise ammonia to asparagine were those which possessed a considerable reserve of carbohydrate; those without this reserve might take up a little ammonia, which would then be stored unchanged. The issue was confused by the fact that certain plants, particularly legumes, also seemed to require calcium, or at any rate the presence of a reagent ($CaCO_3$) in the culture solution to preserve neutrality. Lupins, however (even when provided with this reagent), were plants which failed to produce asparagine, the small amount of ammonia taken up remaining unchanged.

TABLE 36.

(From Prianischnikow, 229.)

10-day etiolated seedlings of *Lupinus luteus*

Culture solution	H_2O	$(NH_4)_2SO_4$	$(NH_4)_2SO_4 +$ $CaCO_3$
	Per 100 seedlings		
	mg.	mg.	mg.
Total-N	567	575	535
Protein-N	152	160	170
Asparagine-N	258	175	158
Ammonia-N	26	57	68

Three types of experiment can readily be suggested to demonstrate the essential nature of the carbohydrate. In the first place, will a seed which contains much carbohydrate reserve in the cotyledons or endosperm behave like lupins if those organs are removed? This was tried with barley seedlings, and the data show clearly that asparagine formation was prevented while ammonia accumulated in unusual amounts.

Secondly, how will lupins behave if given the oppor-

TABLE 37.

(From Prianischnikow, 229.)

Seedlings of *Hordeum sativum* with endosperms removed

Culture solution	H_2O	NH_4Cl	$NH_4Cl +$ $CaCO_3$
	Per 100 seedlings without endosperms		
	mg.	mg.	mg.
Total-N	163	202	242
Protein-N	81	95	87
Asparagine-N	45	57	37
Ammonia-N	4	41	73

tunity of supplying themselves with carbohydrate by means of photosynthesis? The results show pronounced asparagine formation and very little accumulation of ammonia.

TABLE 38.

(From Prianischnikow, 229.)

Culture solution	Seedlings of *Lupinus luteus* grown in light for 15 days		
	H_2O	NH_4Cl	$NH_4Cl + CaCO_3$
	Per 100 seedlings		
	mg.	mg.	mg.
Total-N	885	968	978
Protein-N	617	521	617
Asparagine-N	151	291	227
Ammonia-N	20	35	24

Finally, how will lupins behave when cultured in the dark on ammonium salts if glucose be also added to the nutrient solution? Such an experiment was not easy to perform, but Smirnow (310), working in Prianischnikow's laboratory, developed the necessary technique—a matter of conducting the cultures under sterile conditions —from previous work of Petrow. The answer was clear; asparagine and also protein formation were stimulated, while the accumulation of ammonia was depressed.

TABLE 39.

(From Prianischnikow, 229.)

Etiolated seedlings of *Lupinus luteus*

Culture solution	$(NH_4)_2SO_4 + CaSO_4$	$(NH_4)_2SO_4 + CaSO_4 +$ glucose
	Per 100 seedlings	
	mg.	mg.
Total-N	1,002	1,249
Protein-N	216	380
Asparagine-N	490	619
Ammonia-N	123	82

Prianischnikow summarised all these results in a table which shows the effect of carbohydrate and of light on asparagine formation in seedlings.

Experimental conditions		Results	
Carbohydrate	Light	Asparagine formation	Ammonia enrichment
+	−	+	−
−	−	−	+
+	+	+	−
−	+	−	+

The great significance of ammonia in the nitrogen metabolism of the plant was thus emphasised, and he pointed out that if nitrate absorbed from the soil must first be reduced to ammonia before it becomes available for the synthesis of organic nitrogenous compounds (see p. 111) we may regard ammonia as the α and ω of nitrogen metabolism in that it is the first stage in the synthesis of amino acids and thus of proteins, as well as the end-product of their final decomposition.

These results of Prianischnikow have far-reaching implications for they show that amide formation in the plant requires only ammonia and some non-nitrogenous precursor which *need not be derived from protein at all* but can be provided by glucose or perhaps also by some other nitrogen-free material present. I must therefore again emphasise the statement with which I opened this chapter, that transfer of nitrogen from protein to amides does not necessarily imply that formation of these amides is exclusively a matter of protein metabolism.

CHAPTER V

THE MECHANISM OF AMINO ACID AND PROTEIN SYNTHESIS IN PLANTS

WE have seen (Chapter III) that in the early stages of seedling growth, especially when external sources of nitrogen are not available, a synthesis of protein in the axial organs takes place at the expense of the digestion products of the seed reserves, i.e., to use an expression favoured by earlier workers, it is a *regeneration* from pre-existing protein. The process, however, entails more than a simple transfer of amino acids from the reserve organs to the centres of new growth, for the composition of the old and new proteins differs, and a synthesis of certain of the requisite amino acids—by secondary reactions which involve the production of ammonia from other digestion products and its condensation with nitrogen-free residues, which may be of non-protein origin (Chapter IV)—must take place to an extent which will be determined not only by the composition of the seed reserves but also by the metabolic activities of the seedlings concerned. At a later stage these reserves become exhausted, and the young plant will then depend to an increasing extent on external sources of nitrogen for the elaboration of new protein. This nitrogen is assimilated through the developing root system and will be available for amino acid synthesis in the form of ammonia, nitrate or (with legumes) products metabolised from atmospheric nitrogen by symbiotic bacteria. The course of this synthesis must vary in certain details according to the form of nitrogen provided.

Synthesis of amino acids when nitrogen is available in the form of ammonia. Recent researches would appear to

SYNTHESIS OF AMINO ACIDS

leave but little doubt that in animal tissues a synthesis of amino acids may take place through the condensation of ammonia and the corresponding α-ketonic acid according to the original suggestion of Knoop and Oesterlin (119).

REACTION 1.

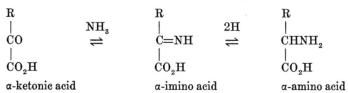

The evidence in favour of glutamic acid synthesis under these conditions is particularly convincing. In the first place, Weil-Malherbe (356) showed that a reversible reductive amination of α-ketoglutaric acid occurred in the presence of brain tissue, and more recently von Euler and his colleagues have investigated the enzyme systems concerned. von Euler et al. (80) found that $l(+)$-glutamic acid was primarily dehydrogenated by codehydrase (Co), on the specific apodehydrase, forming α-iminoglutaric acid and dihydro-codehydrase; the imino acid spontaneously decomposing to α-ketoglutaric acid and ammonia. It was then found easy to synthesise $l(+)$-glutamic acid from the equivalent amounts of the decomposition products with dihydro-cozymase (CoH_2) in the presence of the apodehydrase.

"Glutamic acid dehydrase" is thus an enzyme which synthesises glutamic acid in the organism by reductive fixation of ammonia on to α-ketoglutaric acid, and, at normal physiological values for (H^+), the equilibrium of the reversible reaction is far on the glutamic acid side. In the higher plants, Adler et al. (2) found that the activator was cozymase (codehydrase I), the hydrogen coming into the system by dehydrogenation of triosephosphate, glucose, etc. Furthermore, since the hydrogenated cozymase and apodehydrase are only in dissociated equilibrium, the hydrogen remains quite labile. In confirmation of these findings, we have the evidence of Damodaran and Nair (61) that seedlings of several legumes contain an aerobic dehydrogenase which will convert $l(+)$-glutamic acid to α-ketoglutaric acid and ammonia, while some infiltration experiments described later (see p. 209) show that glutamine can be synthesised from these latter products in blades of perennial rye-grass.

REACTION 2.

$$\begin{array}{c}CO_2H\\|\\CHNH_2\\|\\CH_2\\|\\CH_2\\|\\CO_2H\end{array} + Co \quad \underset{\text{Apodehydrase}}{\rightleftharpoons} \quad \begin{array}{c}CO_2H\\|\\C=NH\\|\\CH_2\\|\\CH_2\\|\\CO_2H\end{array} + CoH_2$$

$$\begin{array}{c}CO_2H\\|\\C=NH\\|\\CH_2\\|\\CH_2\\|\\CO_2H\end{array} + H_2O \quad \rightleftharpoons \quad \begin{array}{c}CO_2H\\|\\C=O\\|\\CH_2\\|\\CH_2\\|\\CO_2H\end{array} + NH_3$$

It is possible that in the plant the synthesis and deamination of all the protein amino acids can be brought about according to reaction 1, for other specific dehydrases, e.g. for alanine and proline, are known to occur in animal tissues; but the fact that protein synthesis—which implies amino acid synthesis—can take place very readily at the expense of ammonia stored as asparagine or glutamine strongly suggests that the "Umaminierung" mechanism of Braunstein and Kritzmann (23, 24) may also be operative. These workers have recently shown that many animal tissues contain a highly active and reversible enzyme system by means of which the amino group and two hydrogen atoms of α-amino acids can be directly transferred to oxalacetic acid or α-ketoglutaric acid, with the formation of the corresponding α-ketonic acids and aspartic or glutamic acids, respectively. Euler et al. (80) state that this exchange-amination with glutamic acid or α-ketoglutaric acid takes place in the presence of enzyme solutions prepared from various higher plants, while, in keeping with such a transfer, Virtanen and Laine (348) have recently shown that aspartic acid can donate amino nitrogen to α-ketonic acids, such as pyruvic acid, if the two reactants are added to a mass of crushed pea plants.

REACTION 3.

$$\underset{\text{α-amino acid}}{\begin{array}{c} R \\ | \\ CH_2NH_2 \\ | \\ CO_2H \end{array}} + \underset{\text{oxalacetic acid}}{\begin{array}{c} CO_2H \\ | \\ CO \\ | \\ CH_2 \\ | \\ CO_2H \end{array}} \rightleftharpoons \underset{\text{α-ketonic acid}}{\begin{array}{c} R \\ | \\ CO \\ | \\ CO_2H \end{array}} + \underset{l(-)\text{-aspartic acid}}{\begin{array}{c} CO_2H \\ | \\ CHNH_2 \\ | \\ CH_2 \\ | \\ CO_2H \end{array}}$$

It is thus very probable that the ammonia assimilated by the root system may be condensed at once with oxalacetic acid, α-ketoglutaric acid or both of these substances to produce aspartic acid, glutamic acid or both amino acids, respectively, either of which can then donate amino nitrogen to such α-ketonic acids as are available and are required for amino acid, and hence for protein, synthesis. Any excess of aspartic acid or glutamic acid would meanwhile condense with a second molecule of ammonia to give asparagine and glutamine, respectively. Chibnall and Grover (44) have prepared asparaginase from germinating barley, while some experiments of Mothes (179) show that many leaves can synthesise asparagine from ammonia and $l(+)$-aspartic acid.

Krebs (125) has shown that many animal tissues can synthesise glutamine from ammonia and $l(+)$-glutamic

REACTION 4.

$$\begin{array}{c}CO_2H\\|\\CHNH_2\\|\\CH_2\\|\\CO_2H\end{array} \quad \begin{array}{c}+NH_3\\\rightleftharpoons\\-NH_3\end{array} \quad \begin{array}{c}CO_2H\\|\\CHNH_2\\|\\CH_2\\|\\CONH_2\end{array} \quad +H_2O$$

$l(-)$-aspartic acid $\qquad\qquad l(-)$-asparagine

acid, the reaction being endothermic and therefore requiring the provision of energy from other reactions. Extracts from the same tissues will bring about the reverse action. In the plant we have evidence that glutamine can be synthesised from these products in the roots of beet (Vickery *et al.*, 334) and in blades of perennial rye-grass (see table 73).

REACTION 5.

$$\begin{array}{c}CO_2H\\|\\CHNH_2\\|\\CH_2\\|\\CH_2\\|\\CO_2H\end{array} \quad \begin{array}{c}+NH_3\\\rightleftharpoons\\-NH_3\end{array} \quad \begin{array}{c}CO_2H\\|\\CHNH_2\\|\\CH_2\\|\\CH_2\\|\\CONH_2\end{array} \quad +H_2O$$

$l(+)$-glutamic acid $\qquad\qquad l(+)$-glutamine

When amino acid synthesis is taking place, in part, at the expense of asparagine or glutamine, as, for example, in certain stages of seedling growth, it will be clear that a direct donation of amino nitrogen can occur only to a limited extent, since these amides contain only one-half of their nitrogen in this form. In such cases the amide-nitrogen, after enzymic hydrolysis to produce ammonia, might temporarily condense with oxalacetic acid or α-ketoglutaric acid if these be available; alternatively, direct condensation of ammonia according to reaction 1

SYNTHESIS OF AMINO ACIDS

might also take place, and a general scheme for amino acid and hence for protein synthesis in plants can be conveniently summarised as follows:

REACTION 6.

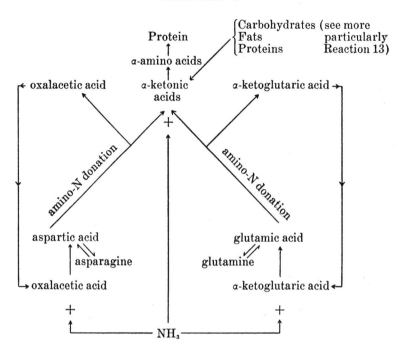

Synthesis of amino acids from products supplied by symbiotic bacteria. Recent research of Virtanen and his colleagues on nitrogen fixation by legume bacteria and excretion of nitrogen compounds from the root nodules (for summary, see 341) has thrown light on the long-standing problem of the form in which legumes receive their nitrogen nutrition from the root nodules, and has incidentally revealed certain of the essential steps in the eventual conversion of atmospheric nitrogen to amino acids.

Inoculated but otherwise sterile cultures of young pea plants in nitrogen-free sand were shown to excrete aspartic acid, β-alanine—a decarboxylation product of the aspartic acid produced through the agency of the nodule bacteria—and oxime nitrogen (Virtanen and Laine, 342, 344). Young active cultures of *Azotobacter chrooccum* and *Bejerinckii* (345) also produced oxime nitrogen (confirming the earlier observation of Endres, 79), as well as aspartic acid. Suspecting that the oxime might be an intermediary product in the synthesis of aspartic acid *via* hydroxylamine and oxalacetic acid, they suspended the nodular mass, detached from the roots of peas, in a dilute solution of this acid at pH 7.0 (346), and observed a pronounced nitrogen fixation in excess of controls. Confirming evidence was then obtained by the actual isolation of the oxime (349) and, since young pea plants had been shown to contain oxalacetic acid (348), Virtanen and Laine suggested that the nitrogen fixation took place in the following way:

REACTION 7.

$$\text{Atmospheric } N_2 \rightarrow ? \rightarrow NH_2OH + \begin{matrix} CO_2H \\ | \\ CO \\ | \\ CH_2 \\ | \\ CO_2H \end{matrix} \rightarrow \begin{matrix} CO_2H \\ | \\ C=NOH \\ | \\ CH_2 \\ | \\ CO_2H \end{matrix} \rightarrow \begin{matrix} CO_2H \\ | \\ CHNH_2 \\ | \\ CH_2 \\ | \\ CO_2H \end{matrix}$$

hydroxylamine   oxalacetic acid   oxime of oxalacetic acid   *l*(—)-aspartic acid

The mechanism by means of which gaseous nitrogen is reduced by the nodule bacteria to hydroxylamine is not clear, but the latter substance condenses with oxalacetic acid received from the host plant to form the corresponding oxime, which then undergoes reduction through the agency of the bacteria to give aspartic acid. This acid is then returned to the host plant to serve as nutrient, or is, in part, excreted into the medium surrounding the nodules.

Virtanen and Laine pointed out that, if aspartic acid

SYNTHESIS OF AMINO ACIDS

should be the only primary amino acid formed during nitrogen fixation, as their results suggested, then the synthesis of other amino acids must take place at its expense. In conformity with this view, and also with that of Braunstein and Kritzmann, they have since been able to show that, in the presence of crushed pea plants, aspartic acid can donate amino nitrogen to pyruvic acid with the formation of alanine. Except in the initial stages, therefore, the synthesis of amino acids, and hence of protein, from atmospheric nitrogen through the agency of symbiotic bacteria, does not differ from that which is operative when nitrogen is supplied in the form of ammonia.

Synthesis of amino acids when nitrogen is available in the form of nitrate. The literature dealing with the storage and assimilation of nitrates, and also with factors influencing nitrate nutrition in plants, has been ably reviewed recently by Nightingale (185) and I shall confine the present discussion to the mechanism of nitrate transformation to organic nitrogen. Many speculative suggestions to explain this have been put forward in the past (see Robinson, 251), but there is now ample evidence to show that the first stage is the reduction of nitrate to nitrite (cf. Eckerson, 70). The later stages are still not well understood. Lemoigne *et al.* (134) claim that a detectable amount of hydroxylamine is present in various leaves, and that the extracts of lilac leaves will reduce nitrite to this substance (135). Since Corbet (56) has shown that the microbiological oxidation of ammonia gives considerable amounts of hyponitrous acid, as well as traces of hydroxylamine, it may be that, in plants, nitrate reduction proceeds through the stages nitrate → nitrite → hyponitrous acid → hydroxylamine → ammonia. Amino acids could then be metabolised from the ultimate or penultimate product by the reactions described above, suggesting that whatever the source of nitrogen available to the plant, amino acid synthesis will finally proceed according to reaction 6.

Protein regeneration in seedlings. It will be recalled that

Schulze, in his later years, inclined to the view that protein regeneration in seedlings would take place preferentially through the agency of asparagine or glutamine rather than from amino acids, because he felt that the amides would be more readily available for this purpose. Prianischnikow, however, had preferred a broader generalisation which, while ranking the amino acids equally with the amides as agents for protein synthesis, gave the preference to the amino acids if these were available, and relegated the amides to the position of storage products which could provide ammonia for amino acid synthesis when required. In Chapter III, I showed that the views of these two workers were not necessarily in conflict if the protein regeneration were considered from the standpoint of amino acid requirements, and a little further consideration will show that both are in full harmony with the scheme of mechanisms portrayed in reaction 6. Indeed, the recent demonstration by Braunstein and Kritzmann that aspartic and glutamic acids can donate amino nitrogen to α-ketonic acids serves only to emphasise over again —and in a way that Schulze could not have foreseen—the very ready "availability" of the two amides as agents for protein synthesis.

Moreover, the relationship which the earlier workers postulated between asparagine and "available carbohydrate" in protein synthesis or regeneration, a relationship which, incidentally, Prianischnikow (see p. 79) found difficulty in accepting, receives a reasonable explanation in terms of reaction 6. The function of the carbohydrate must be to provide the necessary series of α-ketonic acids and there must thus be present in plant cells some enzyme system or systems—at present unknown—which enables this to be brought about. The mere presence of any particular group of carbohydrates, such as the monoses, therefore, need not necessarily lead to protein synthesis if asparagine be available, for conditions may be such that one or more of the mechanisms which control the reactions: sugar $\rightarrow$ α-ketonic acids $\rightarrow$

α-amino acids → protein, are acting as a limiting factor. These views constitute a criticism of Paech's hypothesis which I shall enlarge on more fully in Chapter XI.

Protein synthesis in leaves. The above remarks apply with equal force to the question of protein synthesis in leaves since, from whatever source, internal or external, the nitrogen required for this purpose is provided, it is probable that the ultimate synthesis of amino acids, and hence of protein, must take place according to reaction 6. Hence, in the presence of adequate supplies of sugar and nitrogen—conditions fulfilled in numerous experiments carried out in the past with detached leaves floating on nutrients—the controlling factors will be the necessary enzyme systems concerned. It is to be expected, therefore, that the response will depend on many factors such as leaf age, physiological activities of the plant, etc., which we still find very difficult to correlate with chemical mechanisms, and this aspect of the problem is illustrated in a striking way by some recent work of Pearsall and Billimoria (206).

It may not be out of place to mention here that the vacuum infiltration experiments of Björkstén (16), which have occasionally been quoted in recent years as evidence that protein synthesis in wheat seedlings takes place readily at the expense of urea, acetamide, etc., are worthless. Even if we assume that the observed small increases in protein content are valid, and the experimental technique employed is not reassuring, the fact remains that the fate of the infiltered urea, etc., was not determined, and participation in protein synthesis therefore was not established.

Protein synthesis in ripening seeds. This question was first investigated by Emmerling (76, 77, 78), who, in 1880, collected samples of broad bean plants (*Vicia Faba major*) at various stages of growth and subjected the dried material to a detailed analysis that took him 20 years to complete. He expressed his results in terms of percentages of the dry substance and also in grams per 1000 plants, and the curves that illustrate the accumulation of dry substance—nitrogen, amino nitrogen (Sachsse-Kormann) and protein nitrogen—in the leaves, stems, seeds and roots furnish a vivid picture of the growth of the plant and of the migration of organic solids and of nitrogenous substances from the leaves to the seeds as these matured. Furthermore, the interesting part played by the seed pod, as an intermediate storehouse of nitrogenous substances later furnished to the seed, is particularly well shown. Emmerling did not separately estimate the amount of amino acid amides present in his material, but he concluded that these substances would also migrate with the amino acids from the leaves to the seeds. The process of protein formation in these latter organs thus presented certain analogies with protein regeneration in seedlings, and for this reason Schulze considered that a more detailed chemical study would be of value.

Wassilieff (354), working in Schulze's laboratory, analysed the unripe seeds of various legumes, and found asparagine, arginine, histidine, phenylalanine and valine. Wassilieff (355) and also Pfenninger (214) next showed that during the ripening period much of the protein present in the pods was broken down and the products translocated to the developing seeds. Meanwhile, Schulze himself, with Winterstein (301), had been engaged in a more complete investigation of the non-protein nitrogenous substances in the pods and unripe seeds of the pea (*Pisum sativum*). From the pods they could isolate asparagine (in amount equal to nearly 40 per cent of the total non-protein nitrogen) and much smaller amounts of arginine, histidine, leucine and tryptophan, whereas the unripe seeds contained 40 times as much arginine, but only traces of asparagine, glutamine, lysine and tyrosine. In both cases, they considered that many other amino acids were present, but in amount too small for identification.

Schulze (286) summarised this work in 1911. Emmerling's assumption that the products translocated from the leaves for seed protein formation were amides and amino acids was undoubtedly correct, and the analogy with the regeneration of protein in the developing leaflets of seedlings, at the expense of the seed protein reserves, was a close one. Certain of the necessary products might flow directly from the leaves to the ripening seeds, others, on the contrary, would pass in *via* the pods, as Emmerling had suggested.

To obtain more direct evidence as to the nature of the materials being translocated from the leaves, Schulze analysed whole plants of *Vicia sativa, Trifolium pratense* and *Medicago sativa* at the fruiting stage and found that he could isolate only traces of the three protein bases and leucine, whereas as much as 40 per cent of the non-protein nitrogen present might be due to asparagine (or glutamine). He had himself shown that the normal asparagine content of leaves was low (an observation which has been repeatedly confirmed by later workers), and drew the conclusion that the high asparagine content of the whole plant (including roots) was due to the fact that this substance was in active translocation through the stems.

Schulze considered that these researches with ripening seeds amply confirmed the general conclusion he had already drawn from his work with seedlings, viz., that asparagine and glutamine were the chief products utilised by the plant for protein regeneration. He would not deny, however, that amino acids could also be used for this purpose, but the fact that certain of them could be found in the ripening seed and not in the pod showed, in his opinion, that they would be utilised very slowly, and the high arginine content in unripe pea seeds he regarded as evidence of synthesis *in situ*.

As I have stated before, I prefer Prianischnikow's broader generalisation on the question of protein regeneration, and I agree with him that there is no reason to suppose that amino acids cannot be translocated in the plant as readily as asparagine or glutamine. The recent work of Maskell and Mason (153, 154) with the cotton plant, moreover, in which regional analyses permitted the calculation of concentration gradients, strongly suggests that it is the amino and residual nitrogen which is transported from the leaves to the boll, and that the concentration of amides in the stems is due to storage of those materials.

In my previous discussion of Schulze's views, I pointed out that as we now have some knowledge of the amino acids *required* for protein regeneration, we can give, in certain instances, a new interpretation to his results. In the case of the pea (*Pisum sativum*), for instance, the leaf and seed proteins are somewhat similar in amino acid composition (see p. 83), consequently, if it be assumed that the leaves provide part of the nitrogenous materials for protein synthesis in the pods, there seems to me no valid reason for his suggestion that an extensive conversion of amino acid to asparagine should take place. On the other hand, with wheat (*Triticum sativum*), in which the major part of the nitrogen uptake from the soil normally takes place in the first few weeks of plant growth, the conditions are very different. Schulze and Winterstein (301) investigated the developing grains at the milk stage and could isolate no asparagine or basic amino acids at all, only a very small amount of leucine. In this case the soluble nitrogen passing into the grain must have been derived from the leaf proteins and there can be no doubt that an extensive interconversion of amino acids, probably according to reaction 6 with the intermediary production of asparagine and glutamine, must have taken place. The proteins of wheat blades have not yet been investigated, but those from other *Gramineae* are known to be so alike in amino acid composition that an analysis of cocksfoot (*Dactylis glomerata*) will serve to illustrate the point.

The fact that Schulze and Winterstein could not isolate any asparagine or glutamine is readily explainable in terms of Paech's hypothesis (see pp. 68–70 and footnote on p. 68). The wheat grain is rich in starch and relatively

SYNTHESIS OF AMINO ACIDS

poor in protein, so that during development the entry of a small amount of "available nitrogen" will be accompanied by a very much larger amount of "active carbohydrate." The limiting factor will thus be the non-protein nitrogenous products, and as soon as these reach a certain concentration (too low to permit direct isolation of the individual products in the presence of so much sugar), they will be immediately used for protein regeneration. A

TABLE 40.
(Figures given are in percentages of protein.)

	Wheat grain Gliadin	Wheat grain Glutenin	Leaf blades of cocksfoot
Ammonia	5.22	4.01	1.1
Arginine	2.91	4.72	8.2
Histidine	1.49	1.76	1.5
Lysine	0.64	1.92	5.3
Tyrosine	3.04	4.25	5.0
Glutamic acid	43.66	25.7	14.3
Aspartic acid	0.8	2.0	7.9
Proline	10.4	6.0	2.8
Leucine	6.62	6.3	14.0
Valine	3.34	1.0	6.0
Alanine	2.0	6.1	4.8
Phenylalanine	2.35	2.75	5.0

TABLE 41.
(Computed from Bishop, 15.)

Development of Plumage-Archer barley grains.

Age in days (Anther emergence = 0)	Per 1000 corns					
	Total weight	Total N	Non-protein-N	Salt-soluble protein-N	Hordein-N	Glutelin-N
	g.	g.	g.	g.	g.	g.
3	14.05	0.216	0.086	0.068	0.011	0.054
7	24.68	0.355	0.119	0.057	0.063	0.119
10	32.92	0.449	0.119	0.116	0.104	0.138
14	39.35	0.548	0.121	0.105	0.163	0.158
17	42.42	0.610	0.118	0.121	0.203	0.167
21	44.81	0.656	0.113	0.122	0.223	0.194
28	46.21	0.701	0.100	0.123	0.253	0.229

very neat proof of this is provided by Bishop's (15) careful analysis of the developing barley grain. The nonprotein nitrogen, which clearly represents the primary products of translocation from the blades, never exceeds a certain maximum value, and, as more and more enters the grain during development, it is used at once for the synthesis, in a regular manner, of the three constituent proteins.

CHAPTER VI

THE PREPARATION OF PROTEINS FROM LEAVES

1. *Introduction.*

IN Chapter II, I drew attention to the fact that the pioneer researches of Ritthausen on the preparation, purification and analysis of seed proteins indirectly enabled Schulze to place the whole question of protein metabolism in seedlings on a firm chemical basis. Ritthausen's work was greatly extended by Osborne, and we now have a fairly adequate knowledge of the various classes of proteins present in seeds, and of their individual amino acid composition and biological value as food. Corresponding research on the other parts, especially the green parts, of plants has, however, lagged far behind, and in the chapter devoted to "The Proteins of Green Plants" in the second edition (1924) of his monograph on vegetable proteins, Osborne could still write as follows:

> Various plants contribute large amounts of protein to the ration of farm animals but practically nothing is known of the chemistry of these proteins, even their proportions not yet being established. It is true that the agricultural chemist states the percentages of protein in green fodders, but his analyses are made by indirect methods founded on assumptions unsupported by satisfactory evidence. A serious gap therefore exists in our knowledge of the chemistry of nutrition which makes it impossible to apply to the practical problems of feeding on the farm what has been learned of the nutritive value of the proteins of seeds as well as of the protein concentrates.
>
> Our present meagre knowledge of the protein constituents of living plants is chiefly due to the difficulties encountered in separating the contents of the cells from the enveloping walls. At-

tempts to grind the fresh leaf and extract the contents of the cells with water result in mixtures that cannot be filtered clear, and consequently appear to present no opportunity to obtain the protein in a state fit for chemical examination. Extracting dried leaves with water yields solutions containing only a part of the total nitrogen, and most of this soluble nitrogen does not belong to protein. Such solvents as are generally used for extracting proteins from animal tissues, or from seeds, fail to dissolve much of the residual nitrogen. In consequence of these facts little chemical evidence has been obtained as to the nature of the compounds containing a large proportion of the nitrogen of any plant. . . . It is too early yet to generalise concerning the proteins of green plants, but it would appear from what is now known that relatively small quantities of albumins, globulins and proteoses are present in the juice, and that most of the protein is unlike any of the native proteins thus far described in its behaviour towards acids and alkalis.

Since Osborne wrote these words in 1923, considerable advances have been made, and we are now in possession of a fair body of evidence concerning the nature and amino acid composition of the proteins which can be extracted from the leaves of certain plants, particularly those of economic importance as forage. This new knowledge, some of which is presented for the first time in the following pages, will satisfy, I hope, the more immediate requirements of the protein and agricultural chemist. But much remains to be learned, for the existing methods often fail to extract proteins in reasonable degree of purity from leaves of many important groups of plants. Moreover, the leaf proteins, in contrast to the better-known reserve proteins of seeds, are an integral part of a complex and carefully controlled mechanism, the protoplasm; consequently, if the plant physiologist is ever to be in a position to emulate Lawrence J. Henderson's brilliant researches on blood and attempt to explain protoplasmic behaviour by considering the protoplast as a physico-chemical system—an ideal to which he has not yet attained—he must know more about the conditions under which the proteins exist *in the living cells*. This is a difficult problem to investigate, but it is one towards which, in my opinion, future research should be directed. Such relevant yet admittedly meagre information as we already possess has been gleaned from studies which have had as their primary object the simple characterisation of the *extracted proteins,* and to present it in its right perspective, and, I hope, to stimulate new lines of thought, it will be necessary for me to describe these studies somewhat extensively. This is the reason, and if need be my apology, for the detailed and somewhat disjointed nature of the subject matter which follows.

2. Review of the Early Literature.

In Chapter I, I was able to point to a simple experiment with vetch seedlings which enabled Piria in 1844 to anticipate many of the later conclusions of Pfeffer, Schulze, etc., on asparagine metabolism. In the present instance I can, likewise, call attention to the work of Rouelle, who, as long ago as 1773, prepared proteins from leaves by methods which differ only in detail from those used by both Osborne and Schryver as late as 1919. I quote again from Osborne's monograph (189).

Rouelle [253] announced that [the] glutinous matter, which up to that time was known to exist only in the seeds of wheat, was present also in the other parts of plants. This he considered to be the nutritive substance from which the caseous part of milk was derived. He stated that it was insoluble in water and gave rise to the same products as the gluten of flour, and also that it can be changed into a body having the same odour as cheese, as Kessel-Meyer [in 1759] had proved to be the case with wheat gluten.

Later, Rouelle [253] separated this glutinous substance from the juice of hemlock by heating to a moderate temperature and filtering out the coagulum, which had a bright green colour. He also obtained by fractional coagulation a part which contained nearly all the colouring matter and, at a higher temperature, a part which was nearly free from colouring matter. The colouring matter could also be extracted by digesting the coagulum with alcohol. The protein nature of this substance was proved by the products of destructive distillation. Rouelle was, therefore, the first to obtain evidence of the wide distribution of protein substances in the different parts of plants.

This historical separation of the proteins of leaves into two fractions (really into those of the chloroplasts and of the cytoplasm) by fractional heat flocculation is extremely interesting in view of the important paper published only this year (1938) by Menke. Rouelle's work was extended by Fourcroy (84), who described in 1789 the preparation of what was supposedly pure plant albumin from the juices of various parts of many plants. Probably for the reasons summarised in the citation from Osborne's monograph given above, no further researches of note on the preparation and properties of leaf protein were made from this early date until 1919. Meanwhile, Pfeffer, Schulze, etc., had realised, of course, that the proteins of the axial organs of seedlings and of leaves probably differed, not only in their properties, but also in chemical composition from those present in seeds, but from lack of knowledge they could do no more than follow Fourcroy and refer to them as "albumin."

3. *Preliminary Investigations on the Extraction of Proteins from Leaves.*

What can be regarded as the first serious attempt to prepare the leaf proteins in sufficient quantity for analysis was made almost simultaneously in 1919 by Osborne and Wakeman, who worked with spinach, and by Chibnall and Schryver, who used cabbage. Both groups of workers realised that a method which treated whole leaves as a single unit could take no account of possible variations within individual leaf cells, so that it would be difficult to say whether any preparation thus obtained was derived from the physiologically active (protoplasmic) part of the cell contents or whether it was merely a reserve material analogous to e.g. starch. Such questions, however, were thought at the time to be premature and, in both cases, the aim was merely to obtain a product which could be considered sufficiently pure for preliminary chemical analysis and which would give some indication of the type of protein present in leaves. It was felt that such information would be of great agricultural and dietetical interest, even though it might be quite inadequate to help explain the protein metabolism of the green leaf itself.

Osborne and Wakeman (198) found that, by grinding fresh spinach leaves with water in a Nixtamal mill and centrifuging the mash, a green colloidal solution was obtained which was free from disintegrated cellular material. The addition of about 20 per cent of alcohol gave a voluminous precipitate which consisted of protein, together with other constituents of the leaf cells. Part of the latter was removed by washing with alcohol and ether, and the final product, representing 43 per cent of the total leaf nitrogen, contained 14 per cent of nitrogen. Chibnall and Schryver (49, 50) worked on similar lines, but the leaves were ground with ether-water (to act as a cytolytic agent) and the green colloidal liquid was expressed through muslin in a tincture press. When this liquid was heated to about 80° the crude protein was coagulated. The yields were much lower than those obtained by Osborne and Wakeman, owing to less efficient grinding of the leaves.

Subsequently, Osborne, Wakeman and Leavenworth (199) obtained large quantities of protein from lucerne (alfalfa) by grinding and pressing out the resulting mash through thick filter cloth in a Buchner press. The yellow-brown juice thus obtained appeared to be free from chlorophyll, and, on the addition of about 20 per cent of alcohol, it gave a voluminous precipitate containing 11 per cent of nitrogen and 12 per cent of ash, comprising a mixture of protein, calcium salts of both phosphoric and organic acids, pigments, etc. Treatment with 75 per cent alcohol containing 0.1 per cent hydrochloric acid gave a product which appeared to consist chiefly of the hydrochloride of a protein, for it contained 14.6 per cent of nitrogen and 3.5 per cent of chlorine. The behaviour of this product towards acids and alkalies was so unlike that of most native proteins as to suggest that it was a combination of protein with some, as yet unidentified, complex. This view was supported by the fact that, when the product was boiled with 60 per cent alcohol containing 0.3 per cent sodium hydroxide, the combination was apparently broken, and the protein, on reprecipitation with acid, now contained 16.3 per cent of nitrogen. The new preparation was readily soluble in a slight excess of either acid or alkali, but mild hydrolysis had obviously occurred during the boiling with alkali, since 10.3 per cent of the nitrogen present in the original product was not precipitated by the acid. Analysis by the Hausmann and Kossel-Patten procedures gave the following results:

	Protein per cent	Nitrogen per cent
Ammonia-N	0.96	5.86
Humin-N	0.60	3.67
Basic-N	3.76	22.98
Non-Basic-N	11.04	67.49
Total nitrogen	16.36	100.00

	Protein per cent	per cent
Tyrosine	3.19	
Histidine	2.56	(containing $N = 0.69$)
Arginine	7.11	(containing $N = 2.29$)
Lysine	3.34	(containing $N = 0.64$)
Total nitrogen of bases		3.62

The press residues, made up of debris of cellular matter, contained about half the total nitrogen of the plant. Cold aqueous sodium hydroxide dissolved a small part of this, boiling 60 per cent alcohol containing 0.3 per cent sodium hydroxide dissolved much more, and the final residue contained only 5.9 per cent of the original leaf nitrogen. The aqueous and alcoholic extracts gave impure protein precipitates on being neutralised with acid, but much of the nitrogen remained in solution, due presumably to hydrolytic changes brought about by the alkali.

Meanwhile, I had been working in Schryver's laboratory with leaves of the runner bean (31). By grinding them with water in an end-runner mill and squeezing the mash through a small bag of cotton lawn, I had been able to separate from the pasty mass of cell-wall debris a green extract heavily charged with protein material. After repeating the operation six times in all, I had been able to separate from 80 per cent to 93 per cent of the total leaf nitrogen, depending on the age of the leaves. Under the microscope, the extracts showed a clear, almost colourless, liquid in which were numerous chloroplast fragments in active Brownian movement. On warming to coagulate the protein all these fragments were removed. The residue of cellular debris, when likewise viewed under the microscope, showed occasional lumps of cell matter with unruptured cells still containing their protoplasm; in addition numerous shreds of broken cells, rarely containing portions of protoplasm, were to be seen. It was on this circumstance—that the cellular matter was not ground to a powder, but was merely ruptured and torn, thus setting free the cell contents—that the success of the method depended, and the completeness of the extraction was clearly determined by the thoroughness of the grinding operations. With small amounts (*circa* 100–200 g.) of fresh leaf material, the yields of protein were excellent (75–90 per cent of the total), but the products were not very pure, containing only 11–12 per cent of nitrogen and much inorganic material. Details of the samples prepared from leaves of various ages are given in table 42, the simple analysis by the Hausmann method suggesting that the general amino acid composition did not vary significantly throughout the life history of the plant.

TABLE 42.
(From Chibnall, 31.)

Showing the apparent constancy in composition of protein preparations made from leaves of the runner bean at intervals throughout the life history of the plant.

Age of plant above ground in days	Dry weight %	Total protein-N (% total dry weight)	Protein-N extracted (% total leaf protein-N)	Protein preparations			
				Total-N %	In percentages of total-N		
					Amide-N	Humin-N	Basic N
14	11.4	3.8	86.0	11.4	7.4	3.5	21.6
21	13.1	3.7	87.5	11.4	6.6	3.5	22.4
26	13.4	3.6	88.2	11.8	6.7	3.3	21.6
33	13.2	3.3	82.1	9.7	6.6	3.6	21.0
46	15.8	3.1	76.8	10.9	6.7	4.2	20.5
60	15.9	3.5	81.4	11.8	6.4	3.4	20.5
81	15.5	3.3	76.8	11.7	6.3	3.6	22.0
37	16.3	3.4	75.1	11.5	6.5	3.8	21.8
143	16.1	2.9	77.0	11.4	6.4	3.8	21.3

It will be seen that these preliminary experiments with spinach, lucerne, cabbage and runner bean, carried out between 1919 and 1923, had merely shown that, given adequate grinding facilities, the major part of the cell contents, including much protein, could be separated from the cell walls, and that the extracts contained protoplasmic (especially chloroplastic) material in colloidal solution; preparations of protein of any reasonable degree of purity, however, had not been obtained without resort to treatment which brought about partial hydrolysis.

A little later, actually at the time when Osborne was writing the 1924 (2d) edition of the monograph, I was working in his laboratory on an entirely new method for preparing proteins from leaves.

Extraction of the minced tissue with saline solution under various conditions had already shown that very little, if any, of the protein present belonged to the globulin or albumin classes, and, as the preparations which we had obtained by simple grinding with water or ether-water, i.e. under conditions which precluded hydrolytic breakdown, were heavily contaminated with other constituents of the leaf cells which, on warming or on the addition of alcohol, were precipitated, together with the protein, my endeavour was to effect some initial separation of these substances *before* extraction of the protein itself.

It is well known that if leaves are enveloped in filter cloth and pressed, very little "juice" is exuded until the pressure becomes very great. The turgid cell protoplasts retain their semi-permeability and, if the external pressure is raised above turgor pressure, they may lose water, but very little solutes (see Dixon and Ball, 68, Chibnall and Grover, 42); only when the pressure becomes very great do they lose part of the vacuole solutes (see Phillis and Mason, 215). To "squeeze" the cell fluids from a leaf at all readily, the semi-permeability of the protoplasts must be changed, preferably destroyed, as, for instance, when leaves are immersed in boiling water. Spinach leaves were treated in this way, and the flaccid material was enveloped in filter cloth and pressed in a Buchner press. This gave a clear brown juice which contained some 40 per cent of the total solids of the original leaves. Addition of about 20 per cent of alcohol gave a voluminous white precipitate which had a high ash content and contained very little nitrogen. This was clearly the main adulterant of the protein preparations described above, and suggested that if this clear brown juice could be extracted from the leaves under conditions which did not cause denaturation or coagulation of the proteins, then the press residues of cellular material might, under appropriate treatment, yield protein preparations much purer than we had hitherto secured.

The problem was therefore one of cell permeability, and a procedure applicable to large quantities of leaves similar to that of Dixon and Atkins (67), who cytolysed small quantities of leaves by immersion in liquid air, was sought. Now the anaesthetic action of chloroform and ether on plant cells, either as vapour or in aqueous solution, had been known to plant physiologists for many years, the increase in permeability ultimately leading to the death of the cell. But the extreme ease with which these organic liquids themselves can bring about complete cytolysis did not appear to have been recorded, and I shall never forget my astonishment when first I immersed a fresh turgid spinach leaf in ether. It became extremely flaccid in a few seconds, started to "bleed" a clear brown juice, and gentle pressure between the palm of the hand and the fingers was all that was necessary to express more than half the leaf fluid. I then teased out some individual cells from fresh spinach leaves. Under the microscope they were seen to be turgid, the protoplast, in which the individual chloroplasts were visible, lining the inside of the cell wall. A drop or two of ether was then run between the glass slide and the cover-slip. Contraction of the protoplast away from the cell wall to about one-half of its original volume was almost instantaneous, and a clear brown fluid, obviously the content of the vacuole, was exuded from it. The fluid passed freely through the cell wall (Chibnall, 33). Here, then, was a possible method of separating the vacuole and protoplasmic materials of the leaf.

When this method was applied on a large scale, as I shall briefly describe later, the brown press-juice was found to contain but little protein, and on the addition of about 20 per cent of alcohol a voluminous precipitate, similar in composition to that from leaves treated with boiling water, was obtained, suggesting that the leaf residues would be rich in protein. This was found to be the case and, by appropriate treatment, proteins were prepared in moderately good yield (34, 46, 47) from spinach (*Spinacea oleracea*), lucerne (*Medicago sativa,* alfalfa) and *Zea Mays*. It was pointed out at the time that the method adopted really separated the proteins of the leaf into three fractions: (1) a water-soluble protein which appeared to be an integral part of the cytoplasm, (2) a water-soluble (colloidal) fraction containing protein which seemed to be in some loose combination with lipoids and (3) a very small amount of protein which might be regarded as a constituent of the cell vacuoles. Attention was, however, directed almost exclusively to fraction (1), for the nitrogen content of these proteins was high (*circa* 16 per cent) and they were pure enough to satisfy the immediate requirement, namely, a preliminary examination of their physical and chemical properties. Fraction (2) was, nevertheless, by far the most interesting from the physiological standpoint, since it obviously contained all, or nearly all, of the chloroplast material, but at the time I could not be certain that part was not of nuclear or cytoplasmic origin, and, although since that time much evidence concerning its chemical composition has slowly accumulated in my laboratory, I have been disinclined to publish any results until a separate investigation on more cytological lines had at least shown that my surmise—that it was essentially chloroplast material —was not wholly incorrect. The necessary information has fortunately been provided quite recently by Menke, who has been working in Noack's laboratory at Berlin-Dahlem during the past few years on cytological and chemical studies of chloroplasts (160, 161, 162), and I can re-state or re-interpret much of my earlier data more convincingly if I first of all give an account of his work.

4. *Definitions of Terms Used in Subsequent Discussions.*

In the green leaf the individual cell (protoplast) is enclosed by a cellulose wall and, in its simplest form, consists of a peripheral layer of protoplasm (living matter) surrounding a large vacuole or sap cavity. The cytoplasm,* which is the general ground-mass of the protoplasm, is a more or less transparent viscid liquid and in it are imbedded the nucleus, numerous chloroplasts and

* In previous publications I have used Strassburger's term cytoplasm to embrace all components (including chloroplasts) of the protoplasm except the nucleus. Modern cytological terminology (cf. Sharp, 307), which differentiates the chloroplasts, is preferable.

other (minor) inclusions. Chemically, the cytoplasm and its attendant organs are emulsion structures of protein, lipoid, pentosans, etc., while the vacuolar sap is a solution of inorganic salts, sugar, amino acids and other products of relatively low molecular weight, whose osmotic action in the vascular sap maintains the turgidity of the cell. As Phillis and Mason (215) have recently shown, some of these solutes, in selective and varying concentrations, are also present in the continuous phases (the "vitaids" of Lepeschkin, 137) of the *living protoplasm* and as such must undoubtedly be regarded as essential constituents of the latter.

On the death of the cell, the viscosity of the cytoplasm may change, due, according to Lepeschkin, to the disintegration of the "vitaids," and the (fluid) emulsion may pass into what is usually regarded as a gel; some such change may take place also in the nucleus and chloroplasts. As I shall show later, this emulsion or gel can be extracted in a variety of ways from the leaf cells and separated from the vacuolar sap, but in the process the simple solutes originally present in the living protoplasm are partly or wholly removed also. It is with this "washed" emulsion or gel that I am chiefly concerned in the account which follows, and since the material is nothing more than a mixture of the colloidal constituents of dead protoplasm, the cytologist has naturally, but to my regret, no term by which I can correctly refer to it. Yet some word or brief expression to embrace collectively all the products present is necessary for my purpose, and I feel I can do no better than follow the example of those few biologists who have been interested in similar researches, and refer to the emulsion or gel and also its morphological components as protoplasmic, cytoplasmic, nuclear and chloroplastic material, respectively; it being understood that the use of these terms implies that any simple water-soluble constituents of low molecular weight may have been partly or wholly removed during the course of preparation.

5. *Menke's Separation of Cytoplasmic and Chloroplastic Material from Spinach Leaves.*

Menke's (162) method of separating protoplasmic material from leaves was a development of that of Noack (187), who showed that the green aqueous extracts made by grinding spinach leaves with water would deposit chloroplastic material on centrifuging, and relied partly on the observations of Andrews (3) and Küster (130) that cell nuclei are denser than the remaining protoplasm. Briefly, the leaves were first ground with water to extract the cell contents, the mash was passed through filter cloth to remove the cellular residue, and the green extract was centrifuged for a short time to remove nuclear and cell-wall material. When observed under the microscope, the extract thus obtained showed green droplets in strong Brownian movement. Trial experiments showed that the chloroplastic material could be flocculated by adding ammonium sulphate to 25 per cent concentration, and, after the precipitate had been removed by filtration, the clear brown filtrate on being warmed to 75° gave a heavy, almost colourless coagulum of cytoplasmic material.

Confirmation that the two products separated in this way represented chloroplastic and cytoplasmic material, respectively, was obtained by following the effect in the living cell under the microscope (dark-field illumination). With this concentration of ammonium sulphate, the chloroplasts were the first part to be coagulated, and only very slowly was the cytoplasm affected, the differentiation being especially marked in the case of *Spirogyra,* but observable in the leaf cells of spinach.

The preparations of chloroplastic and cytoplasmic material, after suitable washing to free them from ammonium sulphate, were extracted with ether and ether-alcohol in succession to remove lipoids.* The lipoid-free

* Chibnall and Channon (41) voiced objections to the use of the term "lipoid" in the case of ether-extracts from leaves, which contain chloro-

TABLE 43.
(From Menke, 162.)

Analysis of chloroplastic and cytoplasmic material prepared by various methods from spinach leaves.

Material	How separated	Protein (N × 6.25) %	Lipoid %	Ash %	Residual material %	Chloroplastic cytoplasmic ratio
A. Chloroplastic	25% Ammonium sulphate	53.4	32.7	–	13.9	1.6:1
B. Cytoplasmic	Warming to 75°	91.9	0.5	–	7.6	
C. Chloroplastic	Acidification with HCl to pH 5.8	57.6	29.7	–	12.7	2.1:1
D. Cytoplasmic	Treatment with alcohol	95.2	0.4	–	4.4	
E. Chloroplastic	Acidification with CO_2 to pH 5.8	54.0	30.4	5.6	10.2	1.8:1
F. Cytoplasmic	Acidification with HCl	85.0	0.7	3.1	11.2	
G. Chloroplastic	Centrifuging	47.7	30.8	17.8	3.7	

residues contained 12.7 per cent and 14.7 per cent of nitrogen, respectively. Further analytical data are given in table 43 (A and B).

In other experiments—for full details of which the original paper must be consulted—the procedure was shortened by non-quantitative extraction of the leaf products, and various reagents were used to precipitate (in succession) the chloroplastic and cytoplasmic material, respectively (C, D, E and F, table 43). The chloroplastic material was also prepared in low yield (about 18 per cent) by centrifuging* (G, table 43).

plastic pigments. The term is used here for simplicity and to avoid confusion in comparing results with those of Menke.

* Menke centrifuged in a cup of 20 cm. diameter rotating at 4000 r.p.m. Using an International Equipment Company high-speed centrifuge at 18,000 r.p.m., I find that the whole of the chloroplastic material is removed in about 3 hours.

It will be seen that in each experiment, practically the whole of the lipoid is found in the chloroplastic fraction, and Menke infers from this result that the cytoplasm contains only a trace of lipoid. With this conclusion I am not in entire agreement. Cytoplasm undoubtedly resembles in some of its properties an oil (lipoid) in water (protein) emulsion, and, on the death of the protoplast through the mechanical injury, this emulsion is broken, with the result that, in the case of spinach leaves, the constituent proteins are "dissolved." At the same time the lipoid, if present to any appreciable extent, would separate as a discrete phase to coalesce into microscopic or sub-microscopic droplets which would exhibit Brownian movement. In my opinion, therefore, it does not necessarily follow that all the particles showing Brownian movement in Menke's leaf-cell plasma were of chloroplastic origin, although all of them would have been carried down in the flocculum of what he regards as chloroplastic material. Nevertheless, I do not doubt that the major part of the lipoid in this fraction is of true chloroplastic origin, otherwise the proportion of protein to lipoid in preparation G, which must have consisted solely of *larger* fragments of these organs, would have been greater than in A, C and E, respectively, and the data in table 43 show, if the differences are significant, the reverse trend. With this reservation, I accept Menke's conclusions and interpret my own results accordingly.

Menke's preparations of protein material were insoluble in water, dilute alkali and dilute acid—due, there is no doubt, to denaturation when lipoids were removed by organic solvents. They were, in fact, similar to those of Osborne and Wakeman (198), which consisted of the whole protoplasmic material prepared by essentially the same method. Following the procedure of the latter workers, Menke was able to obtain purer, but undoubtedly hydrolysed, products by treatment with 60 per cent alcohol containing 0.3 per cent NaOH, and his best cytoplasmic and chloroplastic protein preparations contained 16.3 per cent and 13.8 per cent of nitrogen, respectively.

I shall have occasion to discuss some of these results again later; meanwhile, the chief point I wish to emphasise is Menke's clear demonstration that the droplets (the larger of which appear to be green) that are observed to be in rapid Brownian movement, when cell-free extracts from disintegrated leaves are examined under the microscope, consist (with the reservation stated above) essentially of chloroplastic fragments, and that the cytoplasmic proteins have passed into clear aqueous solution. My own observations with spinach leaves had long since led me to similar conclusions, but without the *ad hoc* evidence now presented by Menke, especially his convincing demonstration with living cells treated with 25 per cent ammonium sulphate, I could only regard them as provisional.

6. *The Ether Method for Preparing Cytoplasmic Proteins from Leaves.*

Reverting to my previous discussion (p. 128) of the ether method for preparing proteins from spinach leaves (33, 34), I should like to give now a new interpretation, based partly on the foregoing work of Menke, of what takes place at various stages in the procedure. The leaves are first cytolysed with ether. On cytolysis, the cell protoplasts collapse to about half their original volume and become freely permeable to the aqueous contents of the vacuoles. The cellulose cell wall is freely permeable, consequently the cells lose their turgidity and become flaccid. The leaves are next enveloped in filter cloth and pressed in the Buchner press. The cells become flattened, but the cellulose wall is not broken, so that the aqueous liquid which is forced through it will contain only those water-soluble constituents of the cell whose molecules are sufficiently small to permit them freely to diffuse through the pores of the cellulose wall, which serves, in fact, as a very convenient ultra-filter. These pores must be of relatively large diameter, for in the case of spinach and lucerne a small amount of water-soluble protein passes through, and the material held back is, at this stage, essentially the colloidal components of the protoplasm. The press extract thus obtained is a clear brown fluid, while the residue within the flattened cells is a green gel. On removal from the press, the flattened leaves readily imbibe water and swell up very nearly to their original volume. By alternate pressing and soaking in water the true water-soluble constituents of the leaf cells can be readily washed away.

At this stage, then, we have a cellular residue containing colloidal protoplasmic material that has been washed free from diffusible substances, and an aqueous extract containing the vacuole solutes, any diffusible material of the protoplasm and, in some instances, the vacuolar protein, as I regard it. The leaf residues are next ground in a meat chopper with water. A certain proportion of the cells are now torn open and the green gel of protoplasmic material is dispersed—the fragments of chloroplasts and nuclear material into colloidal solution, as shown by the fact that under the microscope numerous droplets are observed in rapid Brownian movement (the reservation with regard to cytoplasmic lipoid mentioned earlier, p. 133, applies here also), and the cytoplasmic proteins into what is virtually a true solution—probably brought about by an increase in hydrogen-ion concentration consequent on the manifold increase in volume of solvent and the almost complete absence of inorganic buffers. The cell-wall material is then removed by squeezing the mixture through silk gauze.

In one particular experiment (34), 22.64 kg. of fresh spinach leaves, taken from a farm near New Haven on 18th June, 1924, gave (after treatment with ether) 20.7 litres of press-juice from which, by heating to 85°, 10.7 g. of vacuolar protein containing (ash free) 14.0 per cent nitrogen were obtained. This product represented only 2.4 per cent of the total leaf protein, and, as its amino acid composition differed from that of the cytoplasmic and chloroplastic proteins, I refer to it for convenience as the "vacuolar" protein. There is no evidence, of course, to show whether it was originally present in the vacuolar fluid or had been washed out of the protoplasm. A similar product, but in even smaller yield, was obtained from lucerne (*Medicago sativa*, alfalfa) (46), while every other plant that I have so far examined has given none at all.*

After grinding the leaf residues with water, 25 litres of green colloidal extract were obtained. One half of this was treated directly with acid to flocculate the total protoplasmic material (K, table 44); the other half was filtered through paper pulp, which held back the nuclear† and chloroplastic material, and the clear brown filtrate was treated with acid to flocculate the cytoplasmic material (H, table 44). Both products were then washed with alcohol and ether to remove lipoids.

TABLE 44.

(From Chibnall, 34, and unpublished data.)

Preparation of chloroplastic and cytoplasmic material from spinach leaves by the ether method.

Representing 11.32 kg. (720 g. dry weight) of spinach leaves	H Cytoplasmic material (1)	K Protoplasmic material (2)	L Chloroplastic material (2 − 1)
	g.	*g.*	*g.*
Protein (N × 6.25)	35.0	67.15	32.15
Lipoid	0.7	21.08	20.38
Ash	0.58	14.29	13.71
Residual	nil	14.96	14.96

By difference, the weight of chloroplastic material (L, table 44) can be calculated and hence the data given in table 45, which permit of direct comparison with Menke (table 43).

* Kiesel *et al.* (118) record the presence of a small amount of "vacuolar" protein in the leaves of the watermelon (*Citrullus sp.*) but could find none in the leaves of potato, sugar beet and beet-root.

† I find, in agreement with Menke, that the amount of nuclear material in leaves is extremely small and in the discussions which follow it is not differentiated from the chloroplastic material.

TABLE 45.

Chloroplastic and cytoplasmic material from spinach leaves.

	H Cytoplasmic material	L Chloroplastic material
	%	%
Protein (N × 6.25)	96.5	39.6
Lipoid	1.9	25.1
Ash	1.6	16.9
Residual	nil	18.4

The chloroplastic : cytoplasmic ratio is 2.3 : 1.

Two points may be noted here: (1) the cytoplasmic protein contained (ash free) 16.25 per cent nitrogen and was, chemically, a far purer product than any prepared by Menke or by the earlier methods of Osborne, or of Schryver and Chibnall. This is due, as I stated before, to the removal of contaminating substances in the juice expressed from the leaves immediately following the ether treatment: these substances were then in true aqueous solution, and it appears to me an open question, which I need not discuss further, whether they should be regarded as constituents of the vacuole or of the protoplasm. (2) The chloroplastic material contains a high proportion of ash and unidentified products. The former is definitely due in part to soil particles from the surface of the leaves (high silica content, compare also the high ash of Menke's product G, table 43, obtained in very small yield by centrifuging), and the latter in part to fine fragments of cell material that had passed through the silk gauze used for separating the extract. Menke's products also contain varying amounts of ash and unidentified products, due to different procedures in preparation, and a better comparison of our results can be obtained from a simple protein-lipoid ratio.

These results, taken in conjunction with the fact that in each case the cytoplasmic material was practically free from lipoid, show that Menke's various treatments and the ether-method lead to substantially the same sharp separation of chloroplastic from cytoplasmic material.

TABLE 46.

Protein-lipoid ratios in various preparations of chloroplastic material from spinach leaves.

(In percentages of total protein + lipoid)

	Menke's products (cf. table 43)				Chibnall's product
	A	C	E	G	L
Protein (N × 6.25)	62.0	66.0	64.0	60.8	61.2
Lipoid	38.0	34.0	36.0	39.2	38.8

One other point should be mentioned here. Menke's preparations A and B represent an extraction of over 90 per cent of the protoplasmic gel, G only 15 per cent, while C to F are "non-quantitative." My own, H and K, by the ether method, represent 40 per cent. The unextracted protoplasmic gel is located almost entirely in cells that have remained unopened during the grinding operations, and in my previous publications I have always insisted that there was no reason for supposing that it differed significantly in composition from that extracted from the opened cells. The data given in table 46 support this contention, as do certain amino acid analyses quoted later (table 55).

7. *The Proteins of Spinach Leaves.*

Methods of protein analysis. The particular sample of cytoplasmic material H mentioned in the foregoing section had a high nitrogen content (16.25 per cent, ash free) and consisted essentially of protein. It gave no Molisch test for carbohydrate, no furfural on distillation with hydrochloric acid (Tollens) and was considered at the time to be a pure preparation of one or more proteins. Ten preparations made since then have given very similar products, but the nitrogen content has always been lower (15.3–15.8 per cent, ash free), the Molisch test being faintly positive and the Tollens test giving furfural equivalent to about 2 per cent of pentosan. With other leaves, as I shall show later, even greater variations in nitrogen content have been found (14.4–16.7 per cent), and no amount of purification by a method which would preclude even mild hydrolysis (solution in minimum quantity of alkali, followed by filtration through a thick pad of paper pulp and reprecipitation with the minimum quantity of acid) was successful in removing the impurity that was obviously present. In these preparations, also, an estimation of furfural suggested only 1.5–2.5 per cent pentosan, independent of the actual amount of impurity known to be present. The chemical nature of this impurity long eluded me, but I was convinced that it could not be an integral part of the protein (a prosthetic group), since the amount present in preparations from any particular leaf varied fortuitously (171). It was obviously some highly hydrated, probably colloidal, product of large molecular weight, which followed the protein in and out of solution, and in recent years it has become clear from the work of Bailey and Lugg (see Appendix 1) that it consists of material (e.g. mucilage, pentosan) which gives rise to much furfural on acid hydrolysis. The presence of pentosans in leaves is, of course, well known, and the important role they play in the activities of the plant cell was emphasised strongly by Spoehr (313) and MacDougal (147), who speak (148) of plant protoplasm as "an intermeshed pentosan-protein colloid." My own results bear striking testimony to the truth of this statement.

The partial destruction of proteins and certain amino acids, when these are boiled with strong mineral acid in the presence of pentoses, has been investigated by many workers in the past, but the literature gives no clear indication that under these conditions the Tollens estimation yields practically no furfural at all. This misled me for some years, as most of my protein preparations clearly contained 15–25 per cent of some extraneous material, and, as I have stated before, the Tollens procedure suggested only 1.5–2.5 per cent pentosan. It was not until we had realised that all these protein preparations must be relatively rich in pentosans that we could understand the reason why they all gave, on acid hydrolysis, such large amounts of humin. This humin contains much nitrogen and sulphur, and it arises not only from polymerisation of the furfural, but also from its condensation products formed during the destruction of the liberated cystine (particularly), methionine, tyrosine, tryptophan, arginine and histidine. This finding has necessitated a careful re-examination of the methods to be used in the estimation of these particular amino acids, a brief account of which is given in Appendices 1 and 2. Fortunately, three amino acids, lysine after acid hydrolysis, and both tyrosine and tryptophan after alkaline hydrolysis, can be accurately estimated in protein preparations that contain as much as 200 per cent of extraneous pentosan and hexosan adulterants and are thus available for the characterisation of the protein remaining in leaf residues, or even in leaf material from which all water-soluble non-protein substances have been extracted. I have felt it necessary to call attention here to these analytical difficulties, because all my previous leaf-protein analyses (34, 38, 43, 45, 46, 47, 171, 220) are hereby withdrawn, and in those cases in which sufficient material has been available, new analyses by Lugg and Tristram are now substituted.

Cytoplasmic protein. Cytoplasmic material H (table 44) consisted essentially of protein, although small amounts of both lipoid and pentosan were also present. The lipoid-free material contained 16.25 per cent nitrogen, 1.19 per cent sulphur and 0.13 per cent phosphorus. The last value suggests the possible presence of a small amount of nucleic acid, but deliberate search for the nuclear bases has always given a negative result.

Chloroplastic protein. Preparation K (table 44) represented the whole protoplasmic material. The lipoid-free product contained 11.2 per cent nitrogen and 14.8 per cent ash and the large amount of humin formed on acid hydrolysis suggests that the major part of the other organic material present must be of pentosan and hexosan nature. The proteins present are those of both the cytoplasm and chloroplasts, and the amino acid composition of the latter was obtained by difference. There is a significant variation in the content of lysine (especially) and histidine.

TABLE 47.

Amino acid analyses of the three constituent proteins of spinach leaves.

(Figures given are in percentages of total protein-N.)

	Vacuolar protein	Cytoplasmic protein	Chloroplastic protein
Amide-N	5.8	5.6	5.1
Arginine-N	11.2	14.1	13.9
Histidine-N	2.0	2.2	3.3
Lysine-N	7.3	6.2	4.7
Tyrosine-N	2.6	2.7	2.6
Tryptophan-N	1.4	1.7	1.7
Cystine-N	1.4	1.4	1.2
Methionine-N	1.1	1.3	1.3
Aspartic acid-N		5.5	5.8
Glutamic acid-N		6.5	6.5

8. *Properties and Amino Acid Analyses of the Cytoplasmic Proteins from Leaves of Various Herbaceous Plants.*

By the ether method outlined above, cytoplasmic proteins were obtained in varying yields from the leaves of twelve herbaceous plants, as shown in table 48, the absence of all but a trace of chlorophyll in the alcohol and ether washings of the flocculated preparations showing that no appreciable amount of chloroplast material was present.

Although direct supporting evidence is lacking, I think there can be no doubt that these preparations are really mixtures of many proteins which have similar solubility properties. This is patently so from the fact that they represent the protein fraction of the cytoplasm, itself a

TABLE 48.

(From Chibnall and Grover, 42.)

Details of cytoplasmic proteins prepared from various leaves.

Species	Protein (ash free) N %	Leaf sap pH
Spinacea oleracea (spinach)	16.25	6.57
Vicia Faba (broad bean)	16.77	5.67
Phaseolus multiflorus (scarlet runner)	15.85	6.0
Heracleum Sphondylium (cow parsnip)	15.9	6.19
Medicago sativa (lucerne, alfalfa)	15.76	5.67
Crambe cordifolia	15.7	5.56
Brassica oleracea (cabbage)	14.65	5.6
Zea Mays (maize)	14.4	5.66
Cochlearia Armoracia (horse-radish)	14.37	5.4
Tetragonia expansa (New Zealand spinach)	14.46	6.06
Ficus Carica (fig)	13.4	6.8
Ricinus communis (castor bean)	12.55	–

complex organisation which I have treated, for convenience, as a homogeneous constituent. Furthermore, when kept for any length of time under sterile conditions, either in aqueous solution or in the highly hydrated form following precipitation at the isoelectric point, they undergo slow auto-digestion, showing the presence therein of intracellular enzymes, and hence of foreign protein material. In spite of these facts, however, it is convenient, at the present stage of the investigation, to consider these preparations as chemical entities, and, since they are insoluble in water but soluble in slight excess of either acid or alkali, they can be classified as glutelins.

Amino acid analyses of such preparations as were available have been made recently and are given in table 49. Considering that the proteins were prepared from leaves belonging to five separate plant families, the differences to be observed are strikingly small, even though they are proteins of the same class (cf. the seed globulins), and all of them differ markedly in composition from typical seed glutelins such as the glutenin of wheat flour.

TABLE 49.

(From Lugg, 146b, Tristram, 325, and unpublished data.)

Amino acid analyses of various leaf cytoplasmic proteins.

(Figures given are in percentages of total protein-N.)

Species:	Spinacea oleracea (spinach)	Zea Mays (maize)	Ricinus communis (castor bean)	Cochlearia Armoracia (horse-radish)	Crambe cordi-folia	Phaseolus multi-florus (scarlet runner)
amide-N	5.6	5.4	5.1	–	–	5.4
arginine-N	14.1	14.4	12.9	13.9	–	14.9
histidine-N	2.2	2.1	2.2	1.5	–	2.6
lysine-N	6.2	6.1	6.5	4.95	–	6.1
tyrosine-N	2.7	2.3	2.6	2.7	2.4	2.5
tryptophan-N	1.7	1.6	1.7	1.8	1.6	1.6
cystine-N	1.4	1.1	1.5	1.45	1.2	1.1
methionine-N	1.3	1.3	1.45	1.35	1.2	1.1
aspartic acid-N	5.5	–	5.6	–	–	5.2
glutamic acid-N	6.5	–	6.7	–	–	6.7
Protein (ash free) N, %	16.25	14.4	12.75	14.1	15.7	14.4

Within the limits pH 4.0 and pH 5.0, which can be considered their isoelectric point (or, better, range), the solubility of the proteins, when hydrated, is at a minimum and in the case of spinach is of the order 0.0014 per cent. Below pH 4.0, the proteins are slowly soluble, the solubility increasing with decrease of pH to 2.5, beyond which they are again slowly precipitated. In such an acid solution, the proteins are sensitive to the presence of salts, though not all to the same degree. The protein from spinach is sparingly soluble at about pH 3.0 in the vacuolar sap from the same leaves, whereas those from cabbage and horse-radish will not dissolve in an acid solution containing as little as M/50 sodium chloride. Above pH 5.0, the proteins are readily and increasingly soluble with decrease of (H^+), and are precipitated from solution only by salts such as are used for this purpose in general protein chemistry (ammonium sulphate, etc.), and the effect may here be due to increase of (H^+). Treatment of the precipitated products with alcohol and ether to obtain dry preparations causes denaturation.

The reaction of the leaf extracts from which these proteins were prepared varied from pH 6.87, in the case of the fig, to pH 5.37, in the case of horse-radish, so that the proteins were present in the cytolysed cells and almost certainly also in the living cells, as anions, although at a (H^+) not far removed from that of their isoelectric point.

As components of a physico-chemical system (cf. p. 120), the acid- and base-binding capacities of these proteins are of interest, and they have been calculated in two cases from the known content of basic and dicarboxylic acids. The results are given in table 50, together with those of a typical globulin and albumin.

TABLE 50.

Potential acid- and base-binding capacities of the cytoplasmic proteins.

Amino acids (mols $\times 10^{-5}$) in 1 gr. of protein.

	Edestin	Oval-bumin	Cytoplasmic protein from spinach	Cytoplasmic protein from *Ricinus communis*
Arginine	90.9	32.0	42.8	39.2
Histidine	13.4	9.0	8.9	8.9
Lysine	15.0	34.0	37.6	39.5
Total base	119.3	75.0	89.3	87.6
Glutamic acid	132.9*	95.9*	79.1	81.6
Aspartic acid	82.2*	53.9*	66.6	67.9
Amide-bound acids	129.2	70.0	68.0	61.9
Free dicarboxylic acids	85.9	79.3	77.7	87.6

* New determinations in my laboratory.

CHAPTER VII

THE PROTEINS OF PASTURE PLANTS

1. *Limitations of the Ether Method for Preparing Proteins from Leaves; Recent Researches of Foreman.*

AS I stated in the last chapter, the yield of cytoplasmic protein obtained by the ether method varied from plant to plant, and in a great many cases no protein could be obtained at all. The latter was true of plants of economic importance, such as the forage grasses, in which I have been interested during the past few years, and the problem of having to devise methods to overcome this difficulty has been in part responsible for the slow progress made in these leaf-protein studies. The time has been well spent, however, because the newer methods have given clear aqueous extracts containing protein in a reasonable state of purity which is now known to be derived from both the cytoplasm and the chloroplasts. We have thus for the first time been able to gain some insight into the behaviour of the *chloroplastic proteins in solution*.

The reason for these vicarious yields is that in certain cases the protoplasmic gel obtained after cytolysis, when washed free from vacuole solutes, etc., no longer disperses freely into colloidal solution when the cells are opened by grinding. It must, therefore, be in a different state of aggregation from that appertaining to e.g. spinach leaves. Brief consideration will show that this might be due to the vacuole fluids in the living cell having a very high (H^+), so that after cytolysis, which brings all the cell contents to the same (H^+), that of the protoplasm may have been increased from (say) pH 5.7 to between pH 5 and pH 4, the region within which its constituent proteins are practically insoluble. It is not surprising, therefore, that rhubarb leaves, for example, in which the sap, after cytolysis, is at pH 4.0, should give no cytoplasmic protein at all. But in a large number of cases this simple explanation will not suffice, as can be seen from a study of table 51.

TABLE 51.

(From Chibnall and Grover, 42.)

Comparison of the yield of cytoplasmic protein obtained from various leaves.

Species	pH of sap from cytolysed leaves	Method A %	Method B %
Ficus Carica (fig)	6.85	1.5	–
Spinacea oleracea (spinach)	6.57	14.1	16.7
Heracleum sphondylium (cow parsnip)	6.30	5.1	–
Tetragonia expansa (New Zealand spinach)	6.08	4.5	17.5
Phaseolus multiflorus (scarlet runner)	6.00	16.0	16.4
Vicia Faba (broad bean)	5.67	4.0	–
Medicago sativa (lucerne, alfalfa)	5.67	8.6	15.1
Zea Mays (maize)	5.66	12.0	–
Brassica oleracea (cabbage) "unheaded"	5.60	4.0	15.4
Crambe cordifolia	5.50	6.0	27.0
Cochlearia Armoracia (horse-radish)	5.4–5.5	1.0	19.4

Yet many of these leaves can be shown to contain abundant cytoplasmic protein. If, instead of treating them with ether (Method A, table 51), they are at once minced with water and the green colloidal extract filtered through paper pulp, a clear brown filtrate is obtained which gives a heavy, grey coagulum of cytoplasmic material on heating (Method B). These preparations generally have a lower nitrogen content than those made by Method A, but the proportion of the total leaf-protein nitrogen represented is often very much greater, as shown in table 51. Compare, for instance, the leaves of New Zealand spinach and scarlet runner, both of which give a sap after cytolysis of about pH 6, and are therefore comparable in so far as the effect of alkalinity on the solubility of the proteins

THE PROTEINS OF PASTURE PLANTS 147

is concerned. In the former case, the yield of protein by Method B is four times that of Method A, yet in the latter the yields are equally good by both methods. What is the explanation of these variable yields which in the past have made research on leaf proteins so exasperatingly uncertain? Why do certain leaves, for example, *Wistaria chinensis*, with a sap after cytolysis as alkaline as pH 5.7, give no cytoplasmic protein at all, even by Method B?

On ether cytolysis, the cell protoplasts suddenly contract to about half their original size and vacuolar sap is exuded. This contraction is undoubtedly accompanied by an actual decrease in the volume of the protoplasm and, as the chloroplasts contain such a high proportion of lipoid, the water lost must come in large part from the cytoplasm. The concentration of protein, pentosan, etc., in the latter will accordingly be raised, and there will thus be an increase in viscosity which may give rise to a gel of such rigidity that it does not readily disperse into solution when churned up with water. If, on the other hand, the untreated leaves are passed straight through the mincing machine, the cytoplasm, being still in its original, highly hydrated state, can readily disperse. The different amounts of cytoplasmic protein given by Methods A and B, therefore, may be due to nothing more than the difficulty of effecting solution in the former case. Such a simple explanation, however, will not meet many of the difficulties encountered by those who have attempted to prepare proteins from numerous different types of leaves, and it is for this reason that I think Foreman's recent studies should be given serious consideration.

Foreman (83) has applied the ether method to the preparation of proteins from perennial rye-grass and wild white clover at intervals throughout a season's growth. In keeping with the experience of Miller and Chibnall (171), he found that the yields were low, but they could be improved somewhat by preliminary washing of the cytolysed leaf residues (before the grinding stage) with saturated aqueous ammonium sulphate. The filtered solution of protein ultimately obtained was extremely clear and very nearly colourless. The nitrogen content of the dried and ash-free preparations varied between 11.3 per cent and 14.7 per cent, and an examination of the ash disclosed the fact that part of the phosphorus of the preparations was in organic combination, suggesting the possible presence of nucleoproteins. This result led Foreman to follow the distribution of phosphorus as well as of nitrogen in various extracts separated from the leaves, and unexpected facts were thus revealed which I regard as of possible significance.

For the full details of his procedure and results, reference must be made to the original paper, and I quote here only sufficient data to illustrate the point which I wish to emphasise. Briefly, leaves of perennial rye-grass, after cytolysis with ether, were squeezed as free from vacuolar fluids as possible. The residues were then plunged into boiling water, strained off and again washed thoroughly with boiling water. This last operation removed what I refer to in table 52 as "loosely bound" protoplasmic ni-

TABLE 52.

(Compiled from Foreman, 83.)

Lolium perenne	In percentages of total leaf N			In percentages of total leaf phosphorus		
Date of cut	Vacuole N	"Loosely bound" protoplasmic N	"Firmly bound" protoplasmic N	Vacuole P	"Loosely bound" protoplasmic P	"Firmly bound" protoplasmic P
5 June	12.47	0.01	87.52	44.7	26.6	28.7
11 June	10.15	2.02	87.83	51.8	18.0	30.2
15 June	9.55	4.73	85.72	61.8	11.4	26.9
22 June	8.73	2.25	89.02	55.8	17.1	27.2
30 June	9.56	5.34	85.1	54.6	19.8	25.6
5 July	10.24	2.9	86.86	57.0	16.2	26.8
15 August	11.14	7.2	81.66	52.5	26.1	21.4
24 August	11.34	4.28	84.38	65.7	15.2	19.1
7 September	10.78	2.63	86.59	55.6	25.0	19.4
28 September	11.36	9.01	79.63	44.2	31.3	24.5

trogen and phosphorus, leaving a residue which contained the "firmly bound" protoplasmic nitrogen and phosphorus, respectively.

The noteworthy result is the large and *variable* amount of "loosely bound" protoplasmic phosphorus; furthermore, the large amount of "firmly bound" protoplasmic phosphorus assumes an aspect of equal importance if we consider another observation of Foreman, to the effect that leaves dried at 85° (that is, under conditions in which the protoplasm is coagulated in the presence of an increasing concentration of vacuolar sap) readily yield the *whole of their phosphorus* on simple extraction with either *cold or hot water*. Before passing on to his explanation of these interesting differences, let me call attention to one point which he emphasises, namely, that an analysis of the aqueous extract of dried leaves does not necessarily reflect the composition of the vacuolar fluid in the *living cell*, for about one-half of the phosphorus thus estimated is, as he shows, intimately connected with the protoplasm. The recent researches of Phillis and Mason (215) suggest similar views.

Foreman suggests that the protoplasm of leaf cells contains what he refers to as two types of "greater complexes," which may be distinguished according to their behaviour on changing the medium from vacuolar fluid to plain water. (i) In "complex *a*" the protein is loosely combined with phosphates of potassium or sodium, but only at high (H^+) and in the presence of high enough concentrations of similar phosphates, both of which conditions obtain in the vacuolar fluid. The relative amount of this type of complex present in the protoplasm is reflected in the figures for "loosely bound" protoplasmic phosphorus given in table 52. In the absence of vacuole phosphates and at the lower (H^+) of pH 7—conditions prevailing in the ether method of preparing proteins from leaves—the complex is dissociated into its components, both of which disperse into solution. On adding hydrochloric acid to the isoelectric point, the freed protein will then be precipitated. The content of "complex a" thus determines the yield of "soluble" proteins by the ether method. (ii) "Complexes b," consist of nucleoproteins and protein loosely combined with phosphates of calcium or magnesium, which remain intact and insoluble in the presence of vacuolar sap or of pure water. Foreman considers that the ratio of "complex a" to "complexes b" in the protoplasm will vary with varying ratio in the vacuolar fluids of K + Na to Ca + Mg, either as phosphates, or in such form as will surrender these bases to the phosphate-protein-complex formation. Furthermore, the ratio of the two types of complexes may change abruptly on cytolysis of the cell, due to the penetration of vacuolar fluid with a high concentration of the particular bases concerned. Foreman's own data (table 52) show a varying concentration of total phosphorus and total nitrogen per unit sap volume and also of "loosely bound" protoplasmic phosphorus during growth, suggesting that the amount of protoplasmic protein in the form of "complex a" fluctuates during growth as a reserve or storage, being sometimes broken down and supplying the need for breakdown products at the growing points while re-forming in other circumstances at other times. "Complex a" is thus the labile fraction of the protoplasmic protein, and it is of interest to recall (cf. p. 11) that as early as 1872 Pfeffer had suggested the possibility that protein might actually diffuse from one part of the plant to another in the form of a potassium salt or potassium phosphate complex! Foreman has not specifically differentiated the chloroplasts from the general mass of protoplasm, but it is clear that he provides for the proteins of the former in his "complexes b." And it should be noted that he envisages the possibility that part, or even the whole, of the cytoplasmic protein may be rendered insoluble through a base interchange following cytolysis, and this indeed may well be the true explanation of my failure to obtain cytoplasmic proteins by Method B from e.g. *Wistaria chinensis*.

These deductions of Foreman are of great importance, for the protein-phosphate-salt complexes which he postulates—on reasonable circumstantial evidence—are welcome as affording some grounds for speculative suggestions in any attempts we may make to postulate a rational chemical explanation of protoplasmic behaviour.

2. *The Ether-Water Method for Preparing Mixed Protoplasmic Proteins from Leaves; Chemical Composition of the Chloroplast.*

Difficulties were experienced when an attempt was made in my laboratory to prepare proteins from forage grasses by the ether method. Unlike the leaves with which I had hitherto worked, the blades of grasses have an abundance of fibres which cause difficulty during passage through a meat chopper. I expected, therefore, to obtain fairly small yields, but actually they were quite negligible, and this in spite of the fact that the sap reaction of the cytolysed leaf cells was about pH 5.7. Acting on the hypothesis that the cytolytic action of the ether might in these cases be unsuitable, ether-water, as a milder agent, was tried (Miller and Chibnall, 171). The yields in some cases were quite good, in others still negligible, and it was not until we had realised the essential point in the employment of ether-water, namely, that it should have been used previously at least once to cytolyse an appropriate amount of grass, that we were then able to bring the process completely under control (Chibnall *et al.*, 45).

The employment of what I shall henceforth refer to as "used" ether-water brings about an entirely different extraction of the protoplasmic proteins than was at first realised. In one of our preliminary, and successful, experiments with cocksfoot (*Dactylis glomerata*) and "fresh" ether-water, half the extract (following the spinach procedure) was filtered to give a clear brown filtrate and the other half was flocculated with acid. The preparation of cytoplasmic material contained (ash-free) 14 per cent nitrogen and of the protoplasmic material 12.4 per cent nitrogen, the yields (nitrogen in terms of total protein nitrogen) being 16.2 per cent and 34.3 per cent, respectively—values close to those obtained with spinach. In the majority of cases, however, the yield was very disappointing.

It was a chance observation which showed us that if "used" ether-water be employed, the yield of protein material passing through the pad of filter-pulp is always high, and in many instances very much higher (up to 30 per cent of the total protein nitrogen) than we had hitherto obtained. It was noted at the time that these protein filtrates, though quite clear and free from all visible particles under the microscope, exhibited a far more pronounced Tyndall effect in reflected light than had been usual with proteins extracted by the ether method, suggesting that more lipoid was present. This was confirmed when the dried preparations were extracted with organic solvents, appreciable chlorophyll and lipoid material being thereby removed. What we failed to notice, however, for some time was that as much protein, or very nearly as much, passed through the filter-pulp as could be obtained by direct flocculation—very little protein was therefore remaining on the filter pad, although this retained much lipoid. We were accordingly obtaining in clear aqueous solution *most of the extracted protoplasmic protein, i.e. that from the chloroplasts as well as from the cytoplasm.* There was no doubt, therefore, that the "used" ether-water had in some way altered the relationship between the protein and lipoid of the chloroplasts, and that the former was now able to disperse freely into water. This, at any rate, has been our experience with the samples of forage we have used; but if the subtle difference in behaviour between "used" ether-water on the one hand and both "fresh" ether-water and ether itself on the other hand is due to causes such as those suggested below, then it seems likely that the effects of "used" ether-water may vary, according to conditions, from sample to sample.

When leaves are immersed in "fresh" ether-water, cytolysis takes place more slowly than with ether itself, but the collapse of the protoplast is very pronounced and the vacuole solutes diffuse out into the ether-water. In respect to the "availability" of the cytoplasmic proteins, therefore, we should expect conditions to be somewhat similar in the two cases, and this is in accordance with experience—when the yield with ether alone is low, that with "fresh" ether-water is generally of the same order. If a second batch of leaves be immersed in this "used" ether-water, cytolysis appears to take place at about the same rate as before, and vacuole solutes are again exuded. Two striking differences, however, can be readily observed. In the first place, there is a relatively enormous uptake of the ether-water by the leaves, either into the intracellular spaces (Sen and Blackman, 306a) or into the protoplast before the semi-permeability is completely destroyed. In the second place, a cut section shows no collapse or shrinkage of the protoplast, which remains gorged with water as in the normal untreated cell. We should expect the protoplasmic proteins therefore to be as readily available as in Method B (p. 146) and this is actually in accordance with experience.

More difficult to explain is the change which renders the chloroplastic proteins available. The early work of Pringsheim, of Meyer and of Schimper suggested that the chloroplast consisted of a colourless protoplasmic matrix or stroma in which were embedded numerous green viscous drops or granules—the so-called grana. Since then opinion has been divided as to whether the living chloroplast shows any visible structure at all; certain cytologists hold the view that it is homogeneous and regard the grana as artifacts of incipient disorganization; others, and in particular those who have very recently investigated the problem by modern optical methods, have veered round to the older idea of a granular structure. The present morphological position has been reviewed by Frey-Wyssling (85).

The chemical composition of the chloroplast has also been the subject of much study, chiefly of a microchemical nature, both proteins and lipoids being early recognised as constituents of possible importance. According to Meyer (167, 168), the former are ergastic substances and the latter the components of the grana in which the chlorophyll is dissolved. Liebaldt (138), from studies of the action of aqueous solutions of alcohols, aldehydes, etc., on chloroplasts, concluded that there were two phases: an aqueous phase that swelled in volume when the tissue was treated with these reagents and a green-coloured lipoid phase, the latter being in a microscopic emulsion-like distribution in the former. Under the action of the abovementioned agents, these two phases could be separated, the emulsion becoming at first sub-microscopic and then microscopic. In further development of this view, arguments have been advanced for a fluid nature of the emulsion on the one hand, and for a more solid or even gel structure on the other (cf. Menke, 160). My own observations are in agreement with the gel structure, otherwise the "non-availability" of the chloroplastic proteins by Methods A and B which remains complete even on very fine disintegration of the chloroplasts, could only be explained on a water-in-oil emulsion basis. As I have had occasion to mention before (p. 132), the green extract prepared from fresh spinach leaves by Method B, when centrifuged at 18,000 r.p.m., slowly deposits the chloroplast fragments. When a preparation of chloroplast material separated in this way was again centrifuged for a further period at the same speed, a stage was reached when aqueous liquid no longer exuded. Presumably the chloroplast fragments had now more or less coalesced and all interstitial fluid had been squeezed out, so that the preparation contained no more water than that normally present in the aqueous phase (gel) of the original chloroplasts.

The actual water content at this stage (estimated by drying at 108°) was found to be 55 per cent. If we assume that the dried product (there was insufficient of it for comprehensive analysis) had the same composition as the preparations of chloroplastic material given in tables 45 and 46, the aqueous phase of chloroplasts must have contained about 30 per cent of protein, sufficient, according to my experience, to give a very viscous solution which would readily gel.

THE PROTEINS OF PASTURE PLANTS

Under the action of vacuolar fluid (Method B), or ether (Method A), this gel clearly does not disperse when churned up with water, but it does so readily under the action of "used" ether-water. The reason for this can, at present, only be surmised. "Used" ether-water, owing to its dilution with vacuolar sap, contains less ether than "fresh" ether-water, but it also contains vacuole solutes, albeit in low concentration. Since it is rapidly taken up by the leaves, it is possible that its action is in some way connected with the fact that it comes into direct contact with the *outer* surface of the protoplasts before the cytolysis destroys their semi-permeability and so permits the outward passage of undiluted vacuolar sap. Under this milder treatment, part of the lipoid phase, which may be acting as a protective colloid to the protein gel, might become peptised. Chloroplast lipoids, as I show in Appendix III, contain unique calcium (and, in certain cases, magnesium) salts of phosphatidic acid, and a base exchange to give the (water-soluble) potassium, sodium salts on or before cytolysis of the cell might be responsible for these peptising effects.

Such views, however, are speculative, and I have discussed them at some length, as I have indeed the whole of my own and other workers' experiences with the proteins of leaves, because I feel that further progress in this field—which concerns protoplasmic behaviour—must go hand in hand with studies which treat the cell as a physico-chemical system, and two important components of such a system are the proteins and lipoids.

3. *Properties and Amino Acid Analysis of the Mixed Protoplasmic Proteins Prepared from Various Pasture Plants by the Ether-Water Method.*

Properties. The preparations exhibit the same solubility properties as those prepared from the cytoplasm by the ether method, i.e. they are flocculated between the limits pH 4.0–5.0, are freely soluble in a slight excess of alkali and with less ease in a slight excess of acid. The proteins

TABLE 53.

(From Lugg, 146b, c, Tristram, 325, and unpublished data.)

Analyses of mixed protoplasmic proteins.

(Figures given are in percentages of total protein-N.)

Family:	Gramineae									Leguminosae		
Species:	Cocksfoot (*Dactylis glomerata*)	Perennial rye-grass (*Lolium perenne*)	Italian rye-grass (*Lolium italicum*)	Rough-stalked meadow grass (*Poa trivialis*)	Crested dog's tail (*Cynosurus cristatus*)	Chewings fescue (*Festuca rubra* var. *fallax* [Hack])	Hard fescue (*Festuca duriuscula*)	Timothy (*Phleum pratense*)	Lucerne (Alfalfa, *Medicago sativa*)	Wild white clover (*Trifolium repens*)	Red clover (*Trifolium pratense*)
Date of sample	29.9.36 / 7.6.37	6.11.33	26.6.33	27.6.33 / 4.10.32	12.6.33	29.6.33 / 26.9.32	29.9.32	18.10.32	29.6.33	18.10.33	23.10.33
Yield, % of total leaf protein-N	— / —	21.8	20.5	13.7 / 13.6	—	— / 17.8	16.1	9.7	12.6	28.0	24.5
Total-N (ash-free) %	14.0 / 13.1	13.0	14.1	14.0 / 13.6	14.1	14.4 / 14.2	15.0	13.8	14.4	13.2	12.8
Amide-N	5.3 / 5.0	5.0	4.7	5.1 / 4.8	4.7	5.1 / 4.6	4.8	4.8	5.2	5.7	5.4
Arginine-N	13.9 / 15.5	14.5	13.4	16.4 / 15.3	15.1	14.3 / 14.6	14.0	14.8	15.1	15.4	14.9
Histidine-N	2.3 / 2.3	2.4	2.4	2.3 / 2.1	2.6	1.9 / 2.0	2.4	2.4	2.3	1.6	2.9
Lysine-N	6.2 / 6.0	6.2	5.9	5.8 / 5.4	5.8	5.2 / 5.7	5.8	4.9	7.0	6.6	6.5
Tyrosine-N	2.3 / 2.3	2.2	2.3	2.3 / 2.3	2.2	2.4 / 2.3	—	—	2.8	2.5	2.5
Tryptophan-N	1.8 / 1.8	1.8	1.8	1.7 / 1.7	1.8	1.9 / 1.9	—	—	1.9	1.9	1.7
Cystine-N	1.5 / 1.4	1.3	1.3	1.4 / —	1.5	1.3 / —	—	—	1.3	1.0	1.2
Methionine-N	1.3 / 1.2	1.2	1.4	1.5 / —	1.4	1.4 / —	—	—	1.3	1.2	1.2
Glutamic acid-N	— / 7.8	7.2	—	6.8 / —	—	6.6 / —	—	—	6.4	6.6	—
Aspartic acid-N	— / 4.9	5.4	—	5.2 / —	—	4.9 / —	—	—	5.4	4.7	—

THE PROTEINS OF PASTURE PLANTS

of the chloroplast therefore must be of the same general type as those of the cytoplasm.

Amino acid analyses. These are set out in table 53 and as in the case of the cytoplasmic proteins given in table 49 there is again a very pronounced uniformity in composition. This suggests that, as a class, the chloroplastic proteins probably exhibit the same similarity as do those of the cytoplasm, and that, in all the leaves examined, the proportions in which the two classes are present must therefore be fairly constant.

A direct comparison between both types of protein prepared from one particular leaf has not yet been made, but Menke's method of separating chloroplastic material should permit this in future work. Meanwhile, we have the indirect evidence from spinach leaves that they can differ (table 47) and recent work by Lugg suggests that the same may be true of grasses, though he was primarily concerned with attempts to prepare protein samples representative of the whole protein of the leaf, not to effect a separation. His evidence is based on the analysis of products which have been prepared from the same grass by different methods, so that the proportions in which the two types of protein are present is not always the same. These methods, and the major proteins which the preparations should contain, are set forth below.

Method	Protein preparation will contain the under-mentioned components
B Grinding leaf material with sand and water and filtering through paper-pulp	Mainly cytoplasmic protein
C "Used" ether-water	The cytoplasmic protein and most of the chloroplastic protein
D Grinding leaf material with water or mildly alkaline buffer solutions and centrifuging at low speed for a short time	" " "
E As above, but treating with alcohol-ether before centrifuging (1) or filtering (2)	" " "

In the case of perennial rye-grass the variations in tryptophan-N (which can be estimated to about 3 or 4 per cent) are significant, but taken as a whole the data given in table 54 show remarkable constancy.

TABLE 54.

(From Lugg, 146c.)

Analysis of proteins prepared by various methods.

Species	Date of sample	Method of preparation	Yield % of total protein	Total-N %	In percentages of total protein-N				
					Amide-N	Tyrosine-N	Tryptophan-N	Cystine-N	Methionine-N
Cocksfoot	29.9.36	C	39.4	14.1	5.3	2.34	1.82	3.2	
"	7.11.35*	D	28	13.0	5.1	2.32	1.82		
"	7.6.37	D (with water)	60	12.6	5.0	2.39	1.88	3.0	
"	7.6.37	C	25	13.1	5.0	2.32	1.88	3.0	
Perennial rye-grass	6.11.33	C	22	12.9	5.0	2.26	1.74		
"	9.6.37	C	24	13.5	4.7	2.34	1.73	3.3	
"	9.6.37	B	60	12.3	5.1	2.44	2.09	3.1	
"	29.11.37	D (with pH 7 buffer)	33	13.1	5.0	2.43	2.05	3.2	
"	8.6.38	"	46.5	12.5	5.3	2.42	1.97	3.0	
"	8.6.38	E (1)	46.1	13.3	5.2	2.35	1.92	2.9	
"	8.6.38	E (2)	30.2	14.65	5.2	2.39	1.96	2.8	

4. The Total Protein of the Leaf.

The extraction of protein material from leaves by any of the methods which have been discussed above depends ultimately on grinding operations which open the leaf cells and scour out a part or the whole of their protoplasmic contents. The efficacy of the grinding unfortunately varies enormously; with young, non-fibrous leaves, the final debris of cellular material may contain only 15 to 20

* Prepared in Adelaide, Australia, from local cocksfoot.
† From total organic sulphur (see Lugg, Appendix I, p. 272).

per cent of the original leaf protein (table 42), but with others, especially those of economic importance, the amount unextracted may be as high as 85 per cent (table 53). From a nutritive standpoint, the nature and chemical composition of this unextracted protein is of equal importance to that of the isolated product, but the difficulties attending its direct analysis are great.

The debris of cellular material, if examined under the microscope, shows lumps of tissue with numerous intact cells which retain their full complement of protoplasmic gel, and also adhering shreds of opened cells. In previous publications (43, 45), I have stated that I saw no reason to assume that this unextracted protoplasmic gel differed from that which, by a chance action of the grinder, had been scoured out of the cell and thereby dispersed into colloidal solution. I have accordingly regarded the protoplasmic protein extracted from the leaves as being, in bulk, of similar composition to that which remained in the residue. Incomplete, but on the whole satisfactory, analyses have now been obtained which lend support to this contention.

These cellular residues contain from 3 per cent to 6 per cent nitrogen, depending on the initial protein level and the completeness of the extraction. Residues which include the whole protein of the leaf (leaves dried to coagulate protein, the material ground to a powder, extracted with hot water to remove solutes and with organic solvents to remove lipoids), may contain 5 per cent to 8 per cent nitrogen. Both products, regarded as proteins, are grossly impure and, on acid hydrolysis, much destruction of many of the easily estimated amino acids occurs. As is pointed out in Appendix II, lysine is the only amino acid that survives hydrolysis under such conditions unchanged, and then only after preliminary extraction of the protein from the cell material by dilute acid (Tristram, 325). Tyrosine and tryptophan can, however, be estimated to within 3 or 4 per cent after alkaline hydrolysis, and the total organic sulphur can be taken as a measure of the cystine + methionine present (Lugg, Appendix I, pp. 268 et seq.). Including the "amide-N" we thus have five factors with which we can compare the protein of these residues with that isolated in yields of only 15–30 per cent by the "used" ether-water method.

I think these results show that, within the limits to which we can work at the present time, the two sets of products are of similar composition, and that we can regard these "used" ether-water preparations as reasonably representative of the whole protoplasmic protein of the leaf.

5. The Nutritive Value of the Leaf Proteins.

If we accept the validity of the evidence presented in the previous section, we are led to the conclusion that in the case of pasture plants, which appear to contain no

TABLE 55.

(From Lugg, 146b, c, and Tristram, 325.)

Amino acid analysis of the whole protein of dried grass and of a corresponding protein preparation obtained by the ether-water method.

Material	Date	Total-N % total solids	In percentages of total protein-N				
			Amide-N	Ly-sine-N	Tyro-sine-N	Trypto-phan-N	Cystine-N + Methionine-N
Cocksfoot (*Dactylis gomerata*)							
whole grass	29.9.36	7.5	5.1	5.9	2.44	1.90	2.84
whole grass	7.6.37	5.5	5.0		2.53	1.78	3.02
preparation	29.9.36	14.1	5.3	6.2	2.34	1.82	3.17
preparation	7.6.37	13.1	5.0	6.0	2.32	1.80	2.97
Perennial ryegrass (*Lolium perenne*)							
whole grass	29.11.37	5.1	5.2	5.9	2.41	1.78	2.88
whole grass	9.6.37	4.9	5.2		2.48	1.82	2.89
preparation	6.11.33	12.2	5.0	6.2	2.25	1.75	3.25
preparation	9.6.37	13.5	4.7	5.8	2.34	1.75	3.27

TABLE 56.
(Compiled from Lugg, 146b, c, Tristram, 325, and unpublished data.)
Analysis of the protein of various pasture plants.
(Figures given are in percentages of protein.)

Family:	Gramineae									Leguminosae		
Species:	Cocksfoot (*Dactylis glomerata*)	Perennial rye-grass (*Lolium perenne*)	Italian rye-grass (*Lolium italicum*)	Rough-stalked meadow grass (*Poa trivialis*)	Crested dog's tail (*Cynosurus cristatus*)	Chewings fescue (*Festuca rubra* var. *fallax* [Hack])	Hard fescue (*Festuca duriuscula*)	Timothy (*Phleum pratense*)	Lucerne (Alfalfa, *Medicago sativa*)	Wild white clover (*Trifolium repens*)	Red clover (*Trifolium pratense*)
Date of sample	29.9.36 7.6.37	6.11.33	26.6.33	27.6.33 4.10.32	12.6.33	29.6.33 26.9.32	29.9.32	18.10.32	29.6.33	18.10.33	23.10.33
Ammonia	1.08 1.03	1.03	0.97	1.05 1.0	0.97	1.05 0.95	1.0	1.0	1.07	1.17	1.11
Arginine	7.4 8.2	7.6	7.1	8.4 8.1	8.0	7.5 7.7	7.4	7.8	8.0	8.1	7.9
Histidine	1.4 1.5	1.5	1.5	1.5 1.3	1.6	1.2 1.3	1.5	1.5	1.5	1.0	1.8
Lysine	5.5 5.3	5.5	5.2	5.1 4.8	5.1	4.6 5.0	5.1	4.3	6.2	5.8	5.7
Tyrosine	5.0 5.0	4.8	5.0	5.0 5.1	4.8	5.2 5.0	—	—	6.1	5.5	5.5
Tryptophan	2.2 2.2	2.2	2.2	2.1 2.2	2.2	2.4 2.4	—	—	2.4	2.4	2.1
Cystine	2.2 2.0	1.9	1.9	2.0 —	2.2	1.9 —	—	—	1.9	1.4	1.7
Methionine	2.3 2.2	2.2	2.5	2.7 —	2.5	2.5 —	—	—	2.3	2.2	2.2
Glutamic acid	— 14.0	13.0	—	12.2 —	—	11.8 —	—	—	11.4	11.8	—
Aspartic acid	— 7.9	8.8	—	8.4 —	—	7.9 —	—	—	8.8	7.6	—

162 PROTEIN METABOLISM IN THE PLANT

vacuolar protein, protoplasmic proteins prepared by the "used" ether-water method can be regarded as representative of the whole leaf protein available for nutritive purposes.

The amino acid analyses of these products are thus of great agricultural value, and to conform to general usage they are given in table 56 on a weight basis, the data being calculated on the assumption that the mixed proteins contained 17 per cent nitrogen, a point discussed more fully in Appendix I.

Lugg has recently determined the monoamino acids of a cocksfoot preparation by the ester-distillation method; we have therefore a fairly complete analysis of one typical pasture plant protein.

Before attempting to assess the value of these proteins

TABLE 57.

(Compiled from Lugg, 146b and unpublished, Tristram, 325, and Miller, 170.)

Amino acid analysis of cocksfoot protein.

	N % of total protein-N	Weight % of protein (N = 17%)
Ammonia	5.3	1.1
Arginine	15.5	8.2
Histidine	2.3	1.5
Lysine	6.0	5.3
Tyrosine	2.3	5.0
Tryptophan	1.8	2.2
Cystine	1.5	2.2
Methionine	1.3	2.3
Glutamic acid	8.0	14.3
Aspartic acid	4.9	7.9
Glycine	0.4	0.3
Alanine	4.4	4.8
Valine	4.2	6.0
Leucine(s)	8.8	14.0
Phenylalanine	2.5	5.0
Proline	2.0	2.8
	71.2	82.9

for nutritional purposes, it is necessary to make one or two observations concerning the actual samples of leaf material from which they were prepared. All the grasses were taken from manured plots which had been cut at least once a fortnight and the samples consisted almost entirely of blades; the wild white clover was from a luxuriant crop which was cut with a scythe held well above the ground so that the sample consisted chiefly of leaflets from the upper few inches of the plants, whereas the red clover and lucerne were cut about one inch above ground, and the samples consisted of stunted plants with relatively few leaflets. As pasturage, all the samples would have been considered of good quality (20–30 per cent protein).

The variations in amino acid content are surprisingly small, and compare very favourably indeed with those between e.g. the prolamines of various *Gramineae* seeds or the globulins of various *Leguminosae* seeds, so that the digestible protein of a mixed herbage must have a composition approximating fairly closely to that of cocksfoot (table 57). With regard to the indispensable amino acids (Rose, 252), it was to be expected that these would be present in adequate amount, and in so far as they have been actually determined this is found to be true. Tryptophan (2.1–2.4 per cent) and lysine (4.6–6.2 per cent) compare favourably with the corresponding values (1.3 per cent; 6.25 per cent) for casein, while methionine is a little lower (2.2–2.7 per cent), about the average for most seed proteins. In the case of cocksfoot, the value for total leucine is high (the component isomers have not yet been separated), phenylalanine and valine are high, while threonine has yet to be determined. The excellent nutritive value of the leaf proteins is thus very clearly evident.

Incidentally, the high cystine (especially) and methionine content of the leaf proteins may well explain the observation of Nightingale *et al.* (186) that sulphur-deficient tomato plants exhibit the same characteristics as those suffering from nitrogen deficiency, i.e. modification of the protoplasts accompanied by a diminution in the amount of protoplasmic protein present.

It is unnecessary for me to enumerate here the many factors which may be responsible for a fluctuation in the total protein content of green pasturage during the season. I should like to point out, however, that if this change in total protein content were to take place preferentially at the expense of the cytoplasmic protein (or, to use Foreman's nomenclature, "complex a," cf. p. 149)—and there are grounds for assuming that this is the more labile of the two main protein fractions of the leaf (Chapter VIII)—then a seasonal variation in nutritive value of the pasturage could be expected if the cytoplasmic and chloroplastic proteins of the leaves differed significantly in their content of certain of the essential amino acids.

This is a very important question which has been raised recently by Morris, Wright and Fowler (175), who find, from feeding trials conducted with cows, that spring (May) grass, as regards its biological value for milk production, is markedly superior to that of autumn (October) grass. This they ascribe to a difference in nutritive value of the respective grass proteins. The pasture from which their comparison rations of grass were taken had been well manured with nitrogen, phosphorus and potassium, and was kept manured and cut until the autumn feeding was complete, the protein content of the spring grass being about 18 per cent of the dry matter and the autumn grass 20–23 per cent.

The samples of pasture plants used to prepare leaf proteins in my laboratory were grown under very similar conditions to these, and I have collected the analytical data for various spring and autumn products in table 58. In their former studies on the availability of various proteins for milk production, Morris and Wright (173, 174) have stressed the importance of lysine and tryptophan and it is fortunate that the methods employed by Lugg and Tristram permit a protein analysis for these two amino acids in samples of dried fodder.

TABLE 58.

(Compiled from the data of Lugg, 146b and c, and Tristram, 325.)

Comparison of the proteins in spring and autumn samples of pasture grasses.

(Figures given are in percentages of protein.)

Grass	Season	Date	Material	Ammonia	Lysine	Tyrosine	Tryptophan
Cocksfoot (*Dactylis glomerata*)	Spring	7.6.37	Isolated protein	1.0	5.3	5.0	2.2
"	Spring	5.6.32	Whole protein of grass	1.1	5.25	5.2	2.1
"	Spring	7.6.37	"	1.0	—	5.4	2.2
"	Autumn	29.9.36	Isolated protein	1.1	5.5	5.0	2.2
"	Autumn	29.9.36	Whole protein of grass	1.1	5.25	5.2	2.35
Perennial rye-grass (*Lolium perenne*)	Spring	9.6.37	Isolated protein	1.0	5.1	5.0	2.2
"	Spring	8.6.38	"	1.0	—	5.0	2.3
"	Spring	8.6.38	Whole protein of grass	1.0	—	4.9	2.3
"	Autumn	6.11.33	Isolated protein	1.0	5.5	4.8	2.2
"	Autumn	29.11.37	Whole protein of grass	1.0	5.2	5.1	2.2
Rough-stalked meadow grass (*Poa trivialis*)	Spring	27.6.33	Isolated protein	1.0	5.1	5.0	2.1
"	Autumn	4.10.32	"	1.0	4.8	5.1	2.1
Chewings fescue (*Festuca rubra* var. *Fallax* [Hack])	Spring	29.6.33	"	1.0	4.6	5.2	2.4
"	Autumn	26.9.32	"	1.0	5.0	5.0	2.4

The results, admittedly limited in number, do not suggest any really significant seasonal variation in the content of the three amino acids concerned, and the conclusions of Morris, Wright and Fowler need further investigation. The same applies to Crampton's suggestion (58) that the nutritive value of protein in manured grass is higher than that in unmanured, for here again our limited experience with cocksfoot and perennial rye-grass does not suggest any significant variation in lysine, tryptophan or tyrosine with protein level (15–35 per cent).

It may be true, as the above-mentioned workers assume, that the nutritive value of pasturage is governed by the amount of certain essential amino acids which it can provide on digestion in the animal, but it is wrong to consider that these acids must be derived exclusively from protein. Leaves and stems contain considerable amounts of amides, amino acids and other non-protein nitrogenous products, so that certain of these essential amino acids may already be present as such in the pasturage. As will be apparent from the discussions on seedling metabolism given in Chapter III and on leaf metabolism given in Chapter XI, protein synthesis in one part of the plant may necessitate the transfer of amino acids and amides from another, and if certain of these translocation products should be surplus to immediate requirements they may remain stored as such, or be metabolised to amides. In addition, the synthesis of any amino acid may take place either directly, or indirectly, at the expense of ammonia derived from external sources. It follows therefore that the amount of each essential amino acid present as such in the pasturage sap may be wholly unconnected with the actual protein content of the pasturage.

Some recent work of Virtanen and Laine (343) may be quoted in illustration of this point. The total tryptophan content of peas and red clover was determined at various stages of growth, and in both plants the tryptophan nitrogen, in terms of total nitrogen, was at a maximum just before blooming and then decreased rapidly to a value

THE PROTEINS OF PASTURE PLANTS 167

which remained unchanged during subsequent growth. In the later stages of the pea, an actual disappearance of tryptophan was noted. These results led them to suggest that tryptophan plays an important part in the metabolism of the immature plant, possibly as a source of growth-promoting factors (β-indolylacetic acid). This may be true, but it seems to me that in this experiment it was incumbent on Virtanen and Laine to provide evidence that the colorimetric method which they used to estimate tryptophan (that of Winkler, 362, designed for employment with proteins) was not interfered with by other indole derivatives which may also be present in the plant saps—such as β-indolylacetic acid itself.

6. *The Cystine Content of Pasturage and Wool Production.*

This question was raised in 1928 by Robertson (250), who suggested that, since the keratin of wool contained such a high proportion of cystine (about 13 per cent), the carrying capacity of any country for sheep which are grown for wool production might well be determined by the capacity of its pasture plants to produce cystine. Attention then became focussed on the cystine content of grasses and other pasture plants, and the early results obtained were conflicting. The position was reviewed by Pollard and Chibnall (220) in 1934 who concluded, from analysis of their protein preparations, that pasturage should contain ample cystine for the wool protein requirements of the sheep it normally carries. As is pointed out by Dr. Lugg in Appendix I, the procedure used by Pollard and Chibnall to determine cystine in their products was faulty and the results obtained were far too low. Since the provision of fodder containing good quality protein for wool production has recently been stressed by Marston (149), it will be opportune to re-state the position with respect to English pasturage in light of the new analyses given in table 56.

The data can be discussed only in general terms because, in actual practice, the herbage is always mixed and will change considerably in botanical composition during the season. Furthermore, the amount of ingested cystine which would be catabolised or excreted is unknown, so that I can do no more than attempt to show that the total ingested cystine is greater than would be required for wool-protein synthesis.

If we assume that a sheep eats pasturage containing 4 lbs. of dry matter daily, then the 0.39 lb. of cystine necessary for an average fleece of 6 lbs. would be secured during one year if the diet provided 0.0265 per cent of cystine (on the basis of dry matter) for this purpose. Data given in table 56 suggest that a poor pasturage, with a protein level of only 8 per cent, would contain 0.16 per cent of cystine and a good pasturage, with 20–25 per cent of protein, as much as 0.4–0.5 per cent of cystine. These values are far in excess of those required for the fleece which sheep grazing such pastures would be expected to bear.

7. Relative Merits of "True" and "Crude" Protein Applied to Forage Crops.

Finally, before I leave the subject of leaf proteins in agriculture, there is one other point I feel constrained to mention. I am being repeatedly asked by agricultural chemists to express an opinion as to the relative merits of the "true" and "crude" protein values they apply to forage crops, and I have to reply that their question does not permit of any simple straightforward answer; so much depends on the treatment the sample of forage has received after cutting and the purpose for which the protein value is required.

In any planned research undertaken at an agricultural station, the samples of forage will be sent to the laboratory and dried within a few hours of cutting; under these conditions very little protein breakdown can have taken place and the "true" protein value undoubtedly reflects the actual protein content of the living forage. If, however, the samples have been gathered by farmers, and have been dried off slowly in the air or have been sent, without drying, by post to the analyst, then very considerable protein decomposition, with the production of acid amides and amino acids, will have occurred and the "true" protein value will be low. "Crude" protein overcomes all such difficulties, but of course the value includes the non-protein nitrogenous constituents of the forage. Of these, one-third to one-half will normally consist of acid amides and amino acids, and as these have a nutritive value equivalent to protein, a determination of ½ ("true" protein + "crude" protein) which has sometimes been employed in recent years, has much to recommend it in nutritional studies. The other non-protein nitrogenous constituents of the forage are in large part basic substances of unknown constitution (Vickery, 332) and nitrates. The nutritive value of the former is problematical, but we have no evidence that the bacterial flora and protozoa of ruminants cannot make some use of them, while the amount of the latter present in forage is only significant in the early spring bite or immediately following a dressing of nitrogenous fertiliser.

In nutrition studies, therefore, I incline to the belief that the importance of "true" protein can be overemphasised and that in many cases "crude" protein, which is determined with so much less expenditure of time and labour, is of sufficient accuracy for the purpose. It will be obvious, however, from what I have just stated, that each individual case must be considered on its merits, and I can only add that, in studies on plant metabolism, "crude" protein is, in all cases, quite inadmissible. The factor of 6.25 used to convert nitrogen to protein is a little high; a better value would be 6.0 (nitrogen = 16.7 per cent, dry weight), but the difference is small and I hesitate to make any general recommendation.

CHAPTER VIII

PROTEIN METABOLISM IN LEAVES

1. *Introduction.*

WE saw in Chapter VI that the proteins of leaves are integral parts of the cell protoplasm, i.e. of "living" matter; it is to be expected, therefore, that their metabolism, even though it may follow in broad chemical outline that exhibited by the reserve proteins during the development and germination of seeds, will be much more difficult to interpret.

That this is indeed the case is reflected in the many diverse opinions which have been held as to the role of proteins in cellular processes, and it is interesting to note that in recent years certain investigators have put forward suggestions reminiscent of those—long since discarded—which were advanced during the few years immediately following the establishment of the "protoplasm doctrine" by Max Schultze (266) in 1861. Disregarding the "vitalistic" speculations of Pflüger, Detmer, etc. (cf. p. 44), we may recall Borodin's claim that, in all plant cells, the energy changes resulting in respiration were brought about through a continuous breakdown and rebuilding of protein (cf. pp. 31, 32 *et seq.*). This suggestion had not been favourably received by Pfeffer, who pointed out that a continuous decomposition of protein might be possible if the power of regeneration were also present, but that it was essential that the specific constructive elements of the protoplasm, to which it owed its vital powers, should not be irreparably disorganized, however active respiration might be (212). Pfeffer's influence on contemporary opinion, aided no doubt by the reaction which naturally followed the general discrediting of vitalism, soon established the view that the proteins were the more stable units of the protoplasm—the colloidal ground-mass, in fact, on which were superimposed the more labile components—and that these would be decomposed only in the absence of carbohydrate, when, to use Boussingault's old suggestion, they might be "burnt up" in respiration.

This view is still widely held, but numerous researches carried out during the past fifteen years—stimulated by the great advances in micro-methods of chemical analysis —have tended to emphasise once again the essentially "labile" character of the proteins of leaves, so that the possibility of a more or less continuous synthesis and breakdown of these substances as a normal function in leaves is once again being seriously discussed, while Gregory and Sen (102) even go so far as to assume that a "protein cycle" is operative in leaf respiration. It is my purpose to review all this new evidence, and to attempt to assess the validity of the various hypotheses that have been put forward to explain what can best be described as "the regulation of protein metabolism in leaves." As in the case of seedling metabolism, the chemical mechanisms involved demand first consideration and, since those connected with protein synthesis in leaves have already been reviewed in Chapter V, it is necessary to discuss now the question of protein breakdown. In doing this, I shall draw extensively on the results of certain recent researches in which the breakdown has been accentuated by detaching the leaves from the parent plant.

2. *The Non-Protein Nitrogenous Constituents of Leaf Saps.*

The course of protein decomposition in leaves is, at least in the early stages, often much more difficult to follow than that which occurs during the germination of seeds, for leaves contain—in the cell vacuoles and aqueous phases of the protoplasm—numerous relatively simple nitrogenous substances which may themselves take part in the metabolic processes involved. From a chemical standpoint, this complicates the issue, for the presence of sugars and of numerous other non-nitrogenous products such as organic acids in the expressed fluids has made it extremely difficult to identify these non-protein nitrogenous substances, and not only those originating in protein decomposition, but also those normally present in the leaf sap itself.

Earlier workers such as Borodin and Müller, who used the microscope as an aid to identification, failed to detect even asparagine in normal green leaves: Schulze and Bosshard (293), in 1885, were able to isolate this substance, with allantoin, hypoxanthine and guanine, from the leaves of *Platanus orientalis, Acer pseudoplatanus* and *Trifolium pratense,* but, in 1906, Schulze (285) stated that he had never been able to isolate glutamine, or any of the usual amino acids from the green parts of plants unless these had first been kept in the dark for a short period. Glutamine was actually isolated from the leaves of *Vitis vinifera* by Deleano (62) in 1912 and since that time the only investigation of note has been that of Vickery (330), who used the leaves of lucerne (*Medicago sativa*). The sample of protein-free sap employed in his analysis contained 870 g. of organic solids and 55.13 g. of nitrogen, of which 14.4 g. or 26.1 per cent was in the amino-form (as determined by the method of Van Slyke).

Products isolated accounted for only 9.05 per cent of the organic solids, 20.84 per cent of the total nitrogen and about 31 per cent of the amino nitrogen of the sap, illustrating the difficulties of the analysis and how little we know as yet about the simple nitrogenous products that are normally present in the leaf cell vacuoles. In spite of these low yields, Vickery's results are of the greatest importance, for they show that the vacuolar fluids bathing the protoplasm, contain most, if not all, of the various amino acids present in proteins, and further—in striking agreement with the findings of Schulze regarding the sap of germinating seedlings—that the proportions in which they occur may differ widely from those immediately concerned in protein synthesis or decomposition. It should be noted that Vickery (332) has emphasised the relative absence of true protein bases and, since leaf extracts may give large precipitates with phosphotungstic acid, he has issued a strong caution against identifying this as "basic nitrogen" in the usually accepted (protein) sense of the term, for, in large part, it represents substances whose constitution is quite unknown.

3. *A General Survey of Protein Metabolism in Detached Leaves Cultured on Water.*

Although Borodin (21) was unable to identify asparagine in normal green leaves, he had no difficulty in detecting it in various leaves e.g. *Vicia sepium, V. cracca, Lupinus varius spp.* and *Lathyrus odoratus* which had been detached from the plant and kept in a moist atmosphere

TABLE 59.

(From Vickery, 330.)

	Weight of Substance $g.$
Adenine	2.813
Arginine (free)	0.439
Arginine* (combined)	1.343
Lysine (free)	0.448
Lysine* (combined)	2.256
Stachydrin	34.42
Choline	2.767
Trimethylamine	0.305
Betaine	0.228
Asparagine	18.03
Aspartic acid*	6.196
Tyrosine	0.117
Phenylalanine	0.821
Serine	2.407
Leucine	2.070
Valine	2.887
Alanine	1.163
Total	78.709

* Additional, after acid hydrolysis.

for a few days in the dark. As I have pointed out in an earlier chapter, this observation was cited by him as evidence in support of his modification of Pfeffer's hypothesis, in which he claimed that protoplasmic activities in respiration required a continuous breakdown and regeneration of protein, the breakdown giving rise to asparagine, which was immediately used in protein regeneration provided that sufficient carbohydrate (glucose) were available. In its absence, as in detached and darkened leaves, the asparagine would accumulate, and only under such conditions should it become possible to detect it (cf. p. 32).

The quantitative relationship between protein breakdown and asparagine production in green plants was investigated by Schulze and his pupils (293, 296); young oat plants from which the root system had been removed accumulated much asparagine when placed in a dark room with their cut ends in water.

TABLE 60.

(From Schulze and Bosshard, 293.)

Avena sativa	Fresh plants with roots detached	After standing with cut ends in water in the dark for 8 days
Total-N	4.12	4.50
Protein-N	3.51	1.46
Asparagine-N	0.15	1.69
Other soluble-N	0.46	1.35

In the particular case quoted, 74.8 per cent of the lost protein nitrogen had reappeared as asparagine nitrogen. From water-cultured leaves of *Saponaria officinalis* and *Beta vulgaris,* Schulze (276) was able to isolate glutamine, while Kiesel, working in Schulze's laboratory, obtained arginine, histidine, leucine and valine in similar experiments with red clover (117). Since amino acids could not be detected in the plants before treatment, Schulze inferred, although he admitted that his evidence was meagre, that these were the primary products of protein decomposition and that, as in seedlings, they normally underwent a secondary transformation to give asparagine or glutamine.

More direct proof of this was obtained in some experiments of my own in 1924 (37) with leaves of the runner bean, the amino nitrogen being estimated indirectly by Van Slyke's method. Detached leaves were kept with their petioles in water for five days either in strong diffused daylight or in darkness, and, in *both cases,* there was a marked breakdown of protein, with a simultaneous production of much amino nitrogen as well as of acid amides.

TABLE 61.

(From Chibnall, 37.)

Mature leaves of *Phaseolus multiflorus*

Number of days with petioles in water	Protein-N	Non-protein-N	Ammonia-N	Amide-N $\times$ 2	Amino-N, less that due to amides
0	83.50	16.50	0.30	1.78	5.92
5, in light	70.16	29.84	0.37	8.92	11.31
Change	−13.34	+13.34	+0.07	+7.14	+5.39
5, in dark	64.86	35.14	0.57	11.70	9.97
Change	−18.64	+18.64	+0.27	+9.92	+4.05

In percentages of total leaf-N

Two years later, Mothes (176) published the results of a long series of experiments which bring out even more clearly the analogy between protein decomposition in water-cultured leaves and in germinating seeds.

Leaves of the broad bean, when cultured on water in the dark, showed a steady breakdown of protein, and the soluble products thus formed were in large part transformed to amides, ammonia enrichment not being observable until the fifth day, when the leaves had become yellow. Mothes, however, quotes one particular experiment in which very considerable protein decomposition had taken place within twenty-four hours, accompanied by only a very small increase in amides.

TABLE 62.

(From Mothes, 176.)

Detached leaves of *Vicia Faba*	Number of hours with petiole in water	In percentages of total leaf-N			
		Ammonia-N	Amide-N $\times$ 2	Other soluble-N	Protein-N
	0	0.25	2.33	6.33	91.09
	24	0.25	3.78	12.98	82.99

That the primary decomposition products were amino acids, which underwent subsequent oxidation to give ammonia and hence asparagine or glutamine, was postulated from an experiment similar to that of Butkewitsch (table 30, p. 90). The leaves were anaesthetised with chloroform vapour, and the fact that they suffered no permanent injury was shown by their response when exposed later to light and air, for the revived synthetic mechanisms then rapidly converted the accumulated ammonia to amides.

TABLE 63.

(From Mothes, 176.)

Detached leaves of *Phaseolus multiflorus*	In percentages of total leaf-N			
	Ammonia-N	Amide-N $\times$ 2	Other soluble-N	Protein-N
Control	1.0	9.5	7.2	82.3
4 days' anaesthesia	13.6	4.1	19.5	62.8
4 days' anaesthesia, followed by 8 hours in light and then 40 hours in dark	3.1	17.5	17.5	59.5

In the experiment with the broad bean quoted in table 62, the leaves had been detached at the end of a sunny day from plants growing in the open air. Mothes thought there could be no question, therefore, of carbohydrate deficiency and, since rapid protein decomposition had occurred, he could only infer that in detached leaves, as in seedlings, the extent of protein breakdown *in the initial stages* was independent of carbohydrate supply. In the later stages, however, again as in seedlings, the sparing action of carbohydrate could be readily observed. This was shown by experiments in which the leaves were cultured on glucose solution instead of on water, the effect being more pronounced with young leaves and at a lower temperature.

TABLE 64.

(From Mothes, 176.)

Detached leaves of *Phaseolus multiflorus* (young plants)

Number of days with petioles in culture solution	Culture solution	Temperature	In percentage of total leaf-N			
			Ammonia-N	Amide-N $\times$ 2	Other soluble-N	Protein-N
control	–	17°	0.18	4.98	19.89	74.95
4	2½% glucose	17°	0.20	4.95	16.61	77.94
2	water	24°	1.53	7.83	31.11	59.53
2	2½% glucose	24°	0.23	5.73	21.35	72.69

In keeping with the earlier views of Pfeffer and Prianischnikow, Mothes discussed the possibility that the breakdown of protein in detached leaves, and the subsequent oxidation of the amino acids with the ultimate formation of amides, might be due to the supplies of carbohydrate being insufficient to provide the necessary energy for living processes, so that the amino acid residues were being utilised in respiration. But if this were so, he was puzzled to explain my own experiments quoted in table 61, showing very nearly as much protein breakdown in leaves cultured on water in the light—when carbohydrate should not be lacking—as in the dark. A partial explanation of this difficulty was provided by his own experiments, in which he showed that the sparing action of the products of photosynthesis *depended essentially on the age of the leaf.*

TABLE 65.

(From Mothes, 176.)

Detached leaves of Phaseolus multiflorus, with their petioles in water	In percentages of total leaf-N			
	Ammo-nia-N	Amide-N $\times 2$	Other soluble-N	Pro-tein-N
Young leaves				
Control	1.2	1.3	5.2	92.3
8 days in light	1.0	1.8	8.2	89.0
8 days in dark	1.6	10.7	25.0	62.7
Old leaves				
Control	1.5	1.8	6.1	90.6
8 days in light	1.6	12.3	13.5	72.6
8 days in dark	3.0	21.2	14.9	60.9

We meet here for the first time one of the complicating factors that sharply differentiate the problem of protein metabolism in the leaf from that in the seedling, and I shall refer to it again later in my discussions on protein regulation in leaves.

These excellently planned experiments of Mothes undoubtedly clarified the somewhat indefinite position in which the question of protein decomposition in leaves had necessarily been left by Schulze, but do not permit of a more detailed enquiry into the chemical mechanisms involved for two reasons: (i) It is impossible to discuss the protein-sparing action of carbohydrates on protein decomposition when the carbohydrate content of the leaves is left undetermined. The assumptions that leaves exposed to the light must contain sugar (or starch), and that those kept for some days in the dark will be depleted of these substances are not necessarily true; this criticism, I regret to say, is one that can be levelled against much of the published work dealing with leaf protein metabolism. (ii) The work of Chibnall and Sahai, and, more recently, of Michael, shows that, when detached leaves are kept with their petioles in water, there may be a considerable interchange of soluble products between the leaves and petioles.

Chibnall and Sahai (51) kept two samples of mature leaves, with long fleshy petioles, of the Brussels sprout (*Brassica oleracea* v. *bullata*) in the dark with the cut ends in water for four and six days, respectively. They observed that, in the first case, there was a translocation of both sugars and non-protein nitrogenous products from the petioles to the laminae, while, in the second, a similar translocation of sugars occurred, but the movement of the nitrogenous products was in the reverse direction. They found it necessary, therefore, to repeat their experiments—which were concerned with fat-phosphatide-protein balance in leaves—and to analyse the laminae *with the petioles still attached*. This differed from the procedure both of Chibnall and of Mothes in their abovementioned researches on the breakdown of protein in water-cultured leaves, but, since the plants which they used provided leaves with relatively small petioles, it might be thought that the error thus introduced would not be very large. That this is not necessarily true follows from the recent work of Michael.

Michael (169), whose interesting observations on the chlorophyll-protein balance in the leaves of nasturtium (*Tropaeolum majus*) are referred to below, found that cut leaves left in the dark with their petioles in water yellowed much more quickly than those, with or without petioles, which were laid on moist filter paper in a dark room. Suspecting that this might be due to the translocation of protein decomposition products from the leaves to the petioles, he made appropriate analyses, and found that this was indeed the case, the amount of nitrogen lost from the leaf depending on the length of the petiole. In one particular experiment, tests were made with nasturtium leaves bearing a petiole 8 cm. long; analysis showed that the average loss of nitrogen from each leaf was 1.13 mg. or 8.8 per cent of the total, whereas the upper half of the petiole (next to leaf) gained 0.50 mg. and the lower half 0.62 mg. of total nitrogen, respectively. Since of these latter amounts 20 per cent and 28 per cent, respectively, were protein nitrogen, it was clear that congestion of the simple nitrogenous products in the petiole had, by displacement of equilibrium, led to protein synthesis, for it is unlikely that undecomposed protein from the leaf would wander preferentially to the lower end of the petiole.

In the experiments illustrated by tables 61 to 65, the petioles were removed before the leaves were analysed and, since interchange of soluble nitrogenous products between these organs may have taken place during the period of water-culture, it is obvious that the results, which are expressed in terms of total leaf nitrogen, may not accurately reflect the changes that have taken place *in the leaf*. The large decrease in protein nitrogen shown in tables 61, 63 and 65 is undoubted evidence that extensive protein decomposition has occurred, but small changes, especially those in which, on feeding with glucose or exposure to strong sunlight (tables 64, 65), a small increase of protein nitrogen is claimed, need confirmation under more rigidly controlled conditions. To discuss in detail the quantitative changes which occur in water-cultured leaves, it is necessary to turn to other researches in which these uncertainties are not present, but before doing this I feel constrained to refer to the recent and astonishing claim by Fischer (81) that leaves left with their petioles in water can lose up to 50 per cent of their total nitrogen in one day *by diffusion into the water!* He states that this occurs with leaves which have very short (2–3 cm.) petioles of *Impatiens glanduligera* and that bean leaves, with long petioles, may lose as much as 10 per cent. In a quoted experiment with *Syringa,* however, the leaf with petiole lost 0.49 mg. of nitrogen, yet only 0.096 mg. could be recovered from the water. One is tempted to suggest that Fischer's experimental technique, which permits him to compare one single leaf with another, is faulty, for his claim is directly opposed to the cumulative experience of Deleano, Mothes, Chibnall and Michael.

Much more disturbing is the recent demonstration by Pearsall and Billimoria (204, 205) that, under certain conditions, and when inorganic sources of nitrogen are supplied, plant tissues may lose considerable amounts of nitrogen. Cultures of algae such as *Chlorella* on ammonium nitrate showed very large losses when grown in the dark, while young leaves showed the greatest losses when cultured in the light. The following table illustrates the magnitude of the losses observed when leaves of *Narcissus* were floated on three different nutrient solutions in the dark.

TABLE 66.

(From Pearsall and Billimoria, 205.)

In mg. of N per 10 g. leaf tissue

Leaves of *Narcissus pseudonarcissus* Segment	KNO_3		NH_4Cl		Ammonium tartrate	
	N absorbed	N lost	N absorbed	N lost	N absorbed	N lost
1 Base	13.51	3.54	12.08	6.80	13.16	5.23
2	8.23	2.68	10.94	5.63	7.04	1.51
3	9.53	3.75	13.98	5.83	14.13	5.27
4 Apex	12.92	8.02	16.68	13.31	21.79	11.55

The results suggested that the nitrogen loss was in some way dependent upon the amount of nitrogen absorbed and also upon the internal metabolism of the leaf segment as reflected by its age. Under the conditions of the experiment, protein synthesis was confined to the two basal (youngest) segments while the apical ones could show protein hydrolysis. The nitrogen losses, which were similar in all three cases, were, therefore, greater in segments showing protein hydrolysis and presumably depended also on some hydrolysis product. More detailed analyses of the nitrogen taken up showed that, if narcissus or daffodil leaves were initially free from nitrate (as was usual), a significant amount of this was synthesised when ammonium salts were supplied, suggesting that in the scheme given below the nitrate reduction to ammonia is reversible.

As no loss was observed when organic sources of nitrogen, e.g. asparagine, were used, Pearsall and Billimoria conclude that the mechanism of nitrogen loss must involve one of the stages of conversion of inorganic nitrogen. Since Warburg and Negelein (352) had previously shown that *Chlorella* could reduce nitrate to ammonia *via* nitrite, and Eggleton (73) had demonstrated the existence of nitrite in leaves engaged in nitrate reduction (and, incidentally, suggested that nitrogen loss may thus occur in the manner described below), Pearsall and Billimoria put forward the following scheme to explain the production of gaseous nitrogen.

$$HNO_3 \rightleftharpoons HNO_2 \rightleftharpoons NH_3 \rightleftharpoons \text{Amide-N}$$
$$\downarrow \qquad\qquad \updownarrow$$
$$N_2 \text{ lost} \longleftarrow \text{Amino-N} \rightleftharpoons \text{Protein-N}$$

In plant tissues, nitrite may be present as nitrous acid, and, as is well known, Van Slyke (327) has shown that *in vitro* this reacts with amino acids extremely rapidly, whereas, by contrast, its reaction with ammonia or asparagine amide nitrogen is extremely slow. The actual concentration of free nitrous ions in the cells may be low, and must depend on the hydrogen ion concentration of the sap, but we have already seen that this may vary within the cell, as between cytoplasm, cytoplasmic saps and vacuolar sap, respectively, and on this aspect of the question we really have very little *ad hoc* information. The rate of production of gaseous nitrogen, and therefore of nitrogen loss, may actually be extremely slow, but it would be greatest under conditions which provide abundance of amino nitrogen, i.e. in older leaves which tend to protein hydrolysis, or in younger leaves exposed to the light, which synthesise amino acids from ammonia.

These observations of Pearsall and Billimoria are of the greatest importance in studies of plant nutrition, and must also be taken into account in experiments dealing with detached leaves, for there the protein breakdown to give amino acids may lead either to nitrogen losses through interaction with nitrite derived from the nitrate present, or even to the synthesis of nitrate by preliminary oxidation of these amino acids to give ammonia. A synthesis of nitrate in cultured tobacco leaves has indeed been reported by Vickery, Pucher and their colleagues (336, 335), and more recently by M. C. McKee and Lobb (159) in the leaves of Swiss chard and tomato subjected to culture or to air-drying at room temperature.

4. *Decomposition of the Proteins in Detached Leaves.*

In Chapters VI and VII, dealing with the proteins of leaves, I showed that they were to be regarded, at least in very large part, as constituents of the cell protoplasm, and as I wish to give now an extended discussion of the metabolism of these substances in leaves detached from the parent plant, it is pertinent to enquire whether the breakdown of protein, which is always observable within a few hours, takes place at the expense of one or both the two main components of the protoplasm, i.e. whether the proteins of the cytoplasm or of the chloroplasts can be regarded, perhaps only in part, as "reserve proteins" (cf. Nightingale, 185). In such an enquiry we are aided by the fact that the chloroplasts contain pigments which are absent from the cytoplasm, so that by following their gradual disappearance in the detached leaves, we can obtain some measure of chloroplast disintegration.

Sachs (259), in 1863, showed by microscopic observation that in starved leaves the size of the chloroplasts diminished and, in 1883, Zacharias (366) noted that these organs were the parts of the leaf cell richest in protein. Molisch (172) agreed, and, although he thought that the stroma disappeared completely during yellowing, could always obtain positive xanthoproteic tests for protein. Meyer (166) found that the intensity of this reaction varied with the age of the leaf, due, he thought, to the amount of ergastic (non-living reserve) protein present, and noted also that a reduction in the green colour of leaves runs parallel with a reduction in the size of the chloroplasts and in their protein content. These observations were extended by Ullrich (326), who concluded that chloroplasts could be the seat of intense protein synthesis, and that light played no essential role in the process. He also surmised that protein synthesis, unlike carbohydrate synthesis, could also take place in other regions of the leaf cells. All these deductions were made from observations under the microscope, and it is only in recent years that the protein-pigment balance has been studied by chemical methods.

Here the studies of Michael (169) with the nasturtium (*Tropaeolum majus*) are of exceptional interest. He found that detached leaves of this particular plant yellow readily and progressively from the point of convergence of the leaf ribs, with such a sharp boundary between the yellow and green portions that the extent of yellowing could be easily estimated by measurement with a planimeter. The change was strictly a vital process, for leaves with partial yellowing showed no increase in this on being killed.

Visible yellowing could normally not be detected in less than five days, when much protein had already been broken down, but analysis showed that part of the chlorophyll had by then actually disappeared. Comparative analyses were therefore made of the chlorophyll and protein nitrogen contents of leaves detached from the plants at various stages of growth, both before and after being kept in the dark for varying periods. The graph given in figure 3 shows the average of many such results; all the data tend to lie on a straight line, indicating a direct relationship between decrease in chlorophyll and in protein. The line cuts the abscissa above zero at a point which corresponds to the protein level at the time when the completely yellowed leaves die. In experiments covering 4–6 days, the decomposition generally follows a line parallel to the average (4 and 5 in figure 3), for short periods of 1–2 days, however, the line is less steep (1, 2 and 3, figure 3), showing that, in the initial stages, protein is being decomposed more rapidly than chlorophyll, and suggesting preferential utilisation of cytoplasmic proteins.

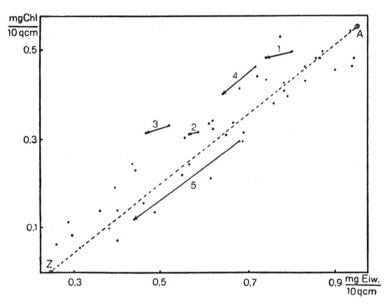

Fig. 3. *Chlorophyll-Protein Ratio in Detached (Half) Leaves during Decomposition in the Dark. Arrows 1–3: one half analysed at once, the other half after 40–48 hours in the dark; arrows 4–5: similar experiments, the other half being kept 4–6 days in the dark. (From Michael, 169.)*

Further support for the hypothesis that, within narrow limits, there is a constant ratio between the content of chlorophyll and of protein was provided by experiments in which protein decomposition was hastened or retarded; in both cases equivalent chlorophyll changes were observed. These experiments, moreover, showed that the proportion of protein decomposition products present was of importance, since yellowing was retarded in young leaves without petioles (no translocation), less so in leaves with petioles (translocation to the latter occurs) and not at all in old leaves (which contain very little nitrogen). With detached autumnal leaves, however, a pronounced difference was noted between those kept in the light and in the dark. The chlorophyll disappeared more slowly in the former case and the yellowing, instead of spreading out evenly from the leaf base, took place slowly over the whole surface of the leaf, which gradually changed from deep green to pale green, yellow green and, finally, pure yellow. The observed protein decomposition, moreover, which was less than in leaves kept in the dark, did not keep pace with the chlorophyll disappearance, suggesting that protein synthesis—at the expense of the decomposition products—was taking place concurrently with the normal protein breakdown. Michael accordingly considered that the difference in the behaviour of the leaves kept in the light was due to the removal of fission products by renewed synthesis under the action of light, so that there was no place of preferential proteolysis (as at the base in darkened leaves) and the leaves yellowed uniformly.

Michael concluded from his experiments that in leaves there is a close relationship between the total protein and the chlorophyll, and, since the primary process of yellowing must be the decomposition of the chloroplast protein-pigment complex, it is clear that a decomposition of the cytoplasmic protein must take place at a similar rate. A mutual relationship between the two main protein groups of the protoplasm is thus postulated.

An earlier experiment of my own with Grover (43) supports this contention of Michael. Leaves of the runner bean were kept with their petioles in water for four days and then treated by Method B (p. 146) to extract proteins and water-soluble nitrogenous products.

In spite of considerable protein decomposition, the relative amounts of chloroplastic and cytoplasmic proteins extracted was the same in both cases, and since we assumed that the unextracted nitrogen represented protein which did not differ materially from that which had been extracted (the validity of this assumption is discussed fully on pp. 159, 160), we concluded that the protein decomposition had taken place at the expense of both the protein fractions concurrently.

TABLE 67.

(From Chibnall and Grover, 43.)

	In percentages of total leaf-N	
Phaseolus multiflorus	Leaves kept with their petioles in water for 4 days	Freshly picked leaves
Chloroplastic protein-N	33.5	35.4
Cytoplasmic protein-N	14.9	16.4
Water-soluble non-protein-N	28.5	15.8
Unextracted-N	23.3	32.4

In further experiments with Jordan (113), again using leaves of the runner bean kept with their petioles in water, it was shown that the breakdown of protein was accompanied not only by a disappearance of chlorophyll, but of carotene, xanthophyll, glycerides and phosphatides as well. Since the ratio total protein : phosphatide was unchanged over a 5-day period, when the leaves were still green, and also over an 8-day period, when they had yellowed, it may be inferred that we are dealing here with a *general breakdown of protoplasm*. Finally, during an investigation of the lipoids of many forage grasses, my colleagues and I have found a fairly constant protein-carotene ratio (confirming Fagan and Ashton, 80a), an increase in protein content due to nitrogenous manuring leading to a corresponding increase in all the various lipoid fractions, and since the whole of the carotene and most of the phosphatide in leaves is present in the chloroplasts, it would appear that there is ample supporting evidence for the contention that a mutual relationship exists between the chloroplastic and cytoplasmic proteins. Synthesis or decomposition of protein in leaves therefore entails approximately equivalent changes in each of these main groups, both of which can thus be regarded—with a lower limit set by the death of the cell—as reserve materials from the standpoint of protein metabolism.

I have, throughout this discussion, been forced, from lack of evidence, to neglect the proteins of the nucleus. Molisch (172) found that the nucleolus and other nuclear material participated only to a very slight extent in the xanthoproteic colour reactions of leaf tissues, and it is probable that the amount of nucleo-protein in leaves is small.

CHAPTER IX

PROTEIN METABOLISM IN LEAVES
(*Continued*)

1. *The Possible Role of Asparagine and Glutamine in the Respiratory Activities of Plants.*

ASPARAGINE and glutamine, as we have seen in Chapter IV, can be synthesised in the plant from ammonia—produced through oxidative deamination of amino acids, or supplied from external sources—and a non-nitrogenous precursor which can be readily metabolised from carbohydrate reserves, especially glucose, and perhaps also from the nitrogen-free residues resulting from the amino acid deamination. In addition, we have the circumstantial evidence brought forward by Prianischnikow (see p. 78), reinforced by deductions which can be drawn from certain experiments of Schulze (see p. 94), that asparagine formation in young plants may be concerned in respiration. To amplify these views it is necessary to discuss at some length the metabolism of the non-nitrogenous precursors of the two amides, for these must clearly be the connecting links between proteins on the one hand and carbohydrates, such as glucose, on the other.

The suggestion that, in the case of asparagine, the non-nitrogenous precursor might be malic acid goes back to 1867 (Beyer, 13), and various C_4-dicarboxylic acids have been assumed to be speculative possibilities even in recent time: Mothes, for instance, as I shall show later, claims to have achieved the synthesis of this amide from the ammonium salts of fumaric, succinic and malic acids. Yet the newer knowledge of amino acid synthesis, summarised in Chapter V, would point to oxalacetic acid as the penultimate precursor of asparagine, and α-ketoglutaric acid as that of glutamine (reaction 6, p. 109): is it possible that in the plant these two ketonic acids can be metabolised—perhaps in processes connected with respiration—from well-known plant acids such as malic acid, succinic acid or citric acid? Recent work on the role of organic acids in animal-tissue respiration presents the plant biochemist with many interesting, if speculative, suggestions.

It has been known now for some years (4, 98, 363) that muscle tissue, for instance, contains enzymes which will bring about the following interactions:

REACTION 8.

$$\begin{array}{c} CO_2H \\ | \\ CH_2 \\ | \\ CH_2 \\ | \\ CO_2H \end{array} \begin{array}{c} -2H \\ \rightleftarrows \\ +2H \end{array} \begin{array}{c} CO_2H \\ | \\ CH \\ || \\ CH \\ | \\ CO_2H \end{array} \begin{array}{c} +H_2O \\ \rightleftarrows \\ -H_2O \end{array} \begin{array}{c} CO_2H \\ | \\ CHOH \\ | \\ CH_2 \\ | \\ CO_2H \end{array} \begin{array}{c} -2H \\ \rightleftarrows \\ +2H \end{array} \begin{array}{c} CO_2H \\ | \\ CO \\ | \\ CH_2 \\ | \\ CO_2H \end{array}$$

succinic acid   fumaric acid   *l*-malic acid   oxalacetic acid

Moreover, Stare and Baumann (315), in keeping with the original suggestion of v. Szent-Györgyi (322, 323), have recently shown that each of these four acids will catalytically promote oxidations, probably of carbohydrates, in minced muscle tissue. In addition, an actual biological interrelationship between fumaric acid and aspartic acid was demonstrated in 1926 by Quastel and Woolf (236a), while Virtanen and Tarnanen (350) have since claimed that weak aspartase activity is exhibited by germinating peas and the leaves of young red clover.

There are thus reasonable grounds for assuming that asparagine might be synthesised in the plant from ammonia and either succinic, malic or fumaric acids, as well as from oxalacetic acid (reaction 6). Such a mechanism, however, would not immediately relate asparagine production with respiration, and a possible connection be-

REACTION 9.

$$\underset{\text{fumaric acid}}{\begin{array}{c}CO_2H\\|\\CH\\||\\CH\\|\\CO_2H\end{array}} \underset{-NH_3}{\overset{+NH_3}{\rightleftarrows}} \underset{l(-)\text{-aspartic acid}}{\begin{array}{c}CO_2H\\|\\CHNH_2\\|\\CH_2\\|\\CO_2H\end{array}} \underset{-NH_3}{\overset{+NH_3}{\rightleftarrows}} \underset{l(-)\text{-asparagine}}{\begin{array}{c}CO_2H\\|\\CHNH_2\\|\\CH_2\\|\\CONH_2\end{array}} + H_2O$$

tween these two activities does not become apparent until we extend our speculation one step further—to the synthesis of glutamine.

Here, another well-known plant acid, citric acid, would appear to take part. In the first place, Krebs and Johnson (126, 127) have shown that this acid has a similar catalytic effect to the above-mentioned C_4-dicarboxylic acids, in promoting the oxidation of carbohydrates in minced muscle tissue. Next, they have shown that when excess of oxalacetic acid is added to minced muscle tissue large amounts of citric acid are formed. They infer that the two extra carbon atoms required for this synthesis arise through the induced oxidation of carbohydrate, which would be in keeping with the demonstration of Knoop and Martius (151) that citric acid is formed *in vitro* if oxalacetate and pyruvate are allowed to react in an alkaline

REACTION 10.

$$\underset{\text{pyruvic acid}}{\begin{array}{c}CO_2H\\|\\CO\\|\\CH_3\end{array}} + \underset{\text{oxalacetic acid}}{\begin{array}{c}CO_2H\\|\\CO\\|\\CH_3\\|\\CO_2H\end{array}} \xrightarrow{+O} \underset{\text{citric acid}}{\begin{array}{c}CO_2H\\|\\CH_2\\|\\HO-C-CO_2H\\|\\CH_2\\|\\CO_2H\end{array}} + CO_2$$

medium and are subsequently treated with hydrogen peroxide.

The oxidation of citric acid itself (by liver dehydrogenase) has been shown by Martius and Knoop (152, 150) to give (in part) α-ketoglutaric acid, and it is of interest to note that the first two intermediary products which they postulate are known plant acids. Wagner-Jauregg and Rauen (351) had previously shown that citric and *iso*citric acids can be dehydrogenated by methylene blue in the presence of various seed extracts, although the products of the reaction were not identified.

REACTION 11.

$$\underset{\text{citric acid}}{\begin{array}{c}CO_2H\\|\\CH_2\\|\\HO-C-CO_2H\\|\\CH_2\\|\\CO_2H\end{array}} \xrightarrow{-H_2O} \underset{\text{aconitic acid}}{\begin{array}{c}CO_2H\\|\\CH\\||\\C-CO_2H\\|\\CH_2\\|\\CO_2H\end{array}} \xrightarrow{+H_2O} \underset{\text{\textit{iso}citric acid}}{\begin{array}{c}CO_2H\\|\\CHOH\\|\\CH-CO_2H\\|\\CH_2\\|\\CO_2H\end{array}} \xrightarrow{-2H} \underset{\text{oxalosuccinic acid}}{\begin{array}{c}CO_2H\\|\\CO\\|\\CH-CO_2H\\|\\CH_2\\|\\CO_2H\end{array}} \rightarrow \underset{\text{α-ketoglutaric acid}}{\begin{array}{c}CO_2H\\|\\CO\\|\\CH_2\\|\\CH_2\\|\\CO_2H\end{array}} +CO_2$$

α-Ketoglutaric acid is readily oxidised in the animal body to succinic acid. Decarboxylation may take place under the action of carboxylase, but Weil-Malherbe (358)

REACTION 12.

$$\underset{\text{α-ketoglutaric acid}}{\begin{array}{c}CO_2H\\|\\CO\\|\\CH_2\\|\\CH_2\\|\\CO_2H\end{array}} \xrightarrow{+O} \underset{\text{succinic acid}}{\begin{array}{c}CO_2H\\|\\CH_2\\|\\CH_2\\|\\CO_2H\end{array}} + CO_2$$

has shown that, under anaerobic conditions, a dismutation may occur, with the parallel production of $l(-)$-α-hydroxyglutaric acid, which is known to occur in plants.

Finally, by reaction 8, the succinic acid passes to oxalacetic acid, and the "citric acid cycle" of Krebs and Johnson is complete. This scheme, they maintain, makes clear the manner in which citric acid and the C_4-dicarboxylic acids act as catalysts; the substances undergo alternate formation and decomposition, and the net effect of the cycle is the oxidation of carbohydrate. If such a cycle should be operative in plants then, under conditions such that ammonia is accumulating in amount likely to become toxic (Prianischnikow), and we know that the tolerance level varies from plant to plant (private communication from Dr. Vickery; see also Burkhart, 25), the free ion may be removed as the ammonium salt of one or more of these acids (Ruhland and Wetzel, 255, 256, 257), α-ketoglutaric acid may be withdrawn from the cycle for the synthesis of glutamine and either fumaric or oxalacetic acids may be withdrawn for the synthesis of asparagine.

Asparagine and glutamine synthesis in the plant can thus be related to the oxidative breakdown of proteins, carbohydrates and fats as shown in reaction 13, page 194.

The amino acids derived from protein digestion undergo oxidation to give ammonia and α-ketonic acids. The latter may condense with oxalacetic acid (step a) to give citric acid and carbon dioxide, or may undergo further oxidation to give succinic acid (step d), (directly, or *via* acetic acid, see p. 97). Fats will also undergo oxidation to give succinic acid in the same direct or indirect way (Verkade, 329). Carbohydrates are oxidised according to the "citric acid cycle," the succinic acid formed from the other two sources entering at the appropriate stage.

The quantitative significance of such a cycle will depend on the rate of the slowest partial step and may thus lead to excess of reactants at certain stages, and it is here that I suggest that the plant cell makes use of asparagine and glutamine. Citric acid and l-malic acid are stored as

194 PROTEIN METABOLISM IN THE PLANT

such in the plant and, unlike oxalic acid, are known to be metabolically active. It is possible that under certain conditions (e.g. extreme starvation), asparagine and glutamine can also be regarded as potential reservoirs of ox-

REACTION 13.

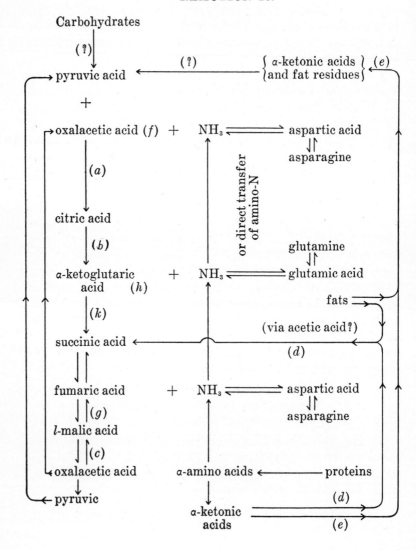

alacetic acid and α-ketoglutaric acid, respectively; thus providing the plant cell with a choice of four sources from which, if need be, it could obtain energy by operating the citric acid cycle, two of them being acidic, the other two neutral and all of them at different energy levels. I infer that the plant, in synthesising asparagine, normally requires oxalacetic acid (reactions 1 and 4), rather than fumaric acid (reaction 9), for such a mechanism would be in harmony with the modern views on amino acid synthesis (reaction 6).

I do not propose to enlarge here on the more general question of carbohydrate respiration in plants or the mechanisms by means of which organic acids may possibly be formed from sugar without participation in respiratory activities. Reaction 13 does, however, emphasise the important part which might be played in plant respiration by certain organic acids, two of which, citric acid and malic acid, are known to be of very widespread occurrence in the vegetable kingdom. Be that as it may, the evidence which has so far been brought forward* to show that organic acids are actively concerned in the metabolic processes of plants, except in a few orders such as the *Crassulaceae,* is extremely meagre, and, as often as not, based on unsatisfactory chemical technique (titration for total acidity). The scheme must therefore be regarded at present as nothing more than a convenient, if speculative, working hypothesis, and in the remainder of this, and in succeeding chapters, I shall attempt to discover the evidence which can be adduced in its favour.

2. *The Nitrogen-Free Precursors of Asparagine and Glutamine in the Plant.*

In the above-mentioned scheme for the intermediary metabolism of asparagine and glutamine in the plant (reaction 13), it was suggested, from evidence based in part on behaviour in other organisms, that the nitrogen-free

* An excellent review was published recently by Bennet-Clark (10).

precursor of glutamine was α-ketoglutaric acid and of asparagine either fumaric acid or oxalacetic acid. Hypotheses formulated in this facile way, however, which at first sight may seem so convincing, may be very difficult to prove: all too often the experiments with plant materials, designed to provide the necessary evidence, lead to conflicting or uninterpretable results. This does not necessarily imply that the hypotheses themselves are false; on the contrary, the failure to obtain the expected result can frequently be shown, on more searching enquiry, to be due to conditions in the experimental plant material being far more complex than was at first assumed to be the case: proteins, carbohydrates and organic acids are all three intimately concerned with the living processes of plants, and the metabolism of one of them cannot be satisfactorily investigated if concomitant changes in either or both of the other two are ignored. This point is emphasised in a striking way by the experiments described below, in which attempts have been made to discover whether certain organic acids can function as nitrogen-free precursors of asparagine.

In the early experiments of Smirnow (310), who cultured seedlings on solutions of the ammonium salts of certain C_4-dicarboxylic acids, the time period of the experiment was long, so that it was sometimes difficult to keep conditions sterile, and, in any case, other reactions connected with growth obscured the issue. These difficulties are to a large extent overcome, however, by the vacuum infiltration method used by Björkstén (16) to promote protein synthesis in leaves. Here the nutrient solution enters the intracellular spaces and is thus brought into intimate contact with a very large number of cells; consequently, under the appropriate conditions, synthesis of amides can take place extremely rapidly.

An interesting series of experiments on these lines was carried out by Mothes (179) in 1933. Preliminary trials showed the need of great care in the treatment of the leaves, otherwise injury to the cells, with a hastening of proteolysis, resulted, and false conclusions could easily be drawn. The infiltered liquid enters the intracellular spaces through the stomata and at any cut ends of tissue, the amount taken up being very great, up to 50 per cent of the fresh weight of the leaf. It is vital that the excess water be removed as soon as possible, and the intracellular spaces refilled with air; this is accomplished through stomatal transpiration, which is rapid under these conditions if the leaves are exposed to diffuse daylight so that the stomata remain open. At the end of 3–4 hours the leaves return to their original weight and the transpiration rate then falls to normal. The leaves are next transferred to a dark chamber and kept in a damp atmosphere (100 per cent humidity) for periods up to 30 hours. Under these conditions, Mothes states that the degree of proteolysis in the infiltered leaves is no greater than in untreated leaves exposed for the same period. The sensitivity of different leaves to the infiltration process varies, and care must be taken to ensure that the nutrient solutions do not cause plasmolysis and exudation of cell sap.

TABLE 68.

(From Mothes, 179.)

Leaves of *Phaseolus multiflorus*	Fresh weight of leaves	NH_3-N infiltered	After 30 hours in dark at 28°	
			NH_3-N	$2 \times$ amide-N
	g.	mg.	mg.	mg.
Analysed at once	24.0	–	0.5	5.5
Infiltered with water	19.2	–	0.7	7.6
Infiltered with ammonium succinate	19.5	25.5	1.1	21.2
Infiltered with ammonium fumarate	20.3	24.1	1.1	28.1
Infiltered with ammonium malate	22.4	24.6	2.1	24.8
Infiltered with ammonium aspartate	23.9	14 (+13.5 amino-N)	1.8	30.4

198 PROTEIN METABOLISM IN THE PLANT

Mothes first infiltered well-nourished leaves of the runner bean with ammonium salts of various organic acids (0.1 M., buffered at pH 6.9–7.0) in an attempt to prove the relationship between aspartic, fumaric, malic and succinic acids embodied in reactions 8 and 9, pp. 190, 191.

After 30 hours, a large production of amide nitrogen had occurred, but Mothes realised that these results did not necessarily prove that the carbon skeleton was provided by the infiltered acid, for this might have come from the reserve material pre-existing in the cell. He therefore repeated the experiment with leaves which had been kept in the dark for five days before infiltration, so as to deplete them as far as possible of available carbohydrate reserves.

TABLE 69.
(From Mothes, 179.)

Phaseolus multiflorus — Per 100 g. fresh leaves

Condition of leaves	Substance infiltered	Fresh weight of sample g.	Total NH_3-N infiltered mg.	NH_3-N infiltered mg.	After 30 hours			
					NH_3-N mg.	2x amide-N mg.	Amino-N less that due to amide-N mg.	Protein-N mg.
Well nourished	Water	22.4	–	–	10.4	30.8	54.2	614
Well nourished	Ammonium malate	21.9	32.0	146	22.3	124.3	83.6	635
Starved (5 days in dark)	Water	23.6	–	–	17.5	60.9	70.1	565
Starved (5 days in dark)	Ammonium malate	23.1	34.9	151	38.5	190.2	82.4	547

The results seemed fairly conclusive, for the starved leaves built up more asparagine than those well nourished, the amount in excess of the control being equal to 0.6 g. per 100 g. of fresh leaves, an astonishingly high figure if it be remembered that the total protein nitrogen liberated was only 18 mg. Mothes quotes many other experiments, especially with *Nicotiana* species, and considered that his results provided good evidence for the aspartic-fumaric-succinic-malic acid relationship set forth above.

At first sight this is indeed so, and the evidence does seem convincing; it is not until we look a little more deeply into the problem that we realise how much Mothes has underestimated its complexity. Protein metabolism—in this case a simple phase of it, amide metabolism—has been inferred from observations confined to changes in nitrogen; the presence or absence, and therefore the effect, of carbohydrates has been deduced from environmental conditions instead of by direct analyses; while the possibility that the necessary organic acid precursors were already present, and available in sufficient amount, in the leaves, has not been considered. Mothes's results, therefore, without further supporting evidence, cannot be accepted as final.

This is the attitude taken by Schwab (304), whose stimulating work on similar lines reopened the whole question, but who has himself not realised, in my opinion, the full complexity of the problem under investigation. Unlike Mothes, who appears to have been careful so to arrange the conditions of his experiments that proteolysis was reduced to a minimum, Schwab in some cases has deliberately employed substances which have led to extensive proteolysis. Moreover, he has designedly infiltered (by centrifuging, instead of *in vacuo*) more dilute solutions of his reacting substances than Mothes, and therefore deals in most cases with relatively small increases in amide nitrogen; otherwise he has followed Mothes's general technique.

Schwab noted first of all that the acetate ion brought about proteolysis in the leaves of the runner bean, and then carried out the following experiment with carbohydrate depleted leaves of *Fittonia Verschaffeltii*.

TABLE 70.
(From Schwab, 304.)

Leaves of *Fittonia Verschaffeltii* Kept in the dark for 4 days (fresh weight 16 g.)	Infiltered with:	After 24 hours in the dark NH_4 succinate	NH_4 acetate	Untreated leaves
		mg.	mg.	
Ammonia-N		1.01	3.13	0.49
Glutamine amide-N		0.67	1.00	0.38
Asparagine amide-N		4.03	4.50	1.47
Protein-N		87.12	75.60	85.12
Increase in *total*-N of amides + ammonia-N		6.22	9.97	–

Total-N infiltered: 7 mg.

The ammonium succinate had induced a strong amide synthesis, and there was no protein decomposition; the acetate-ion, however, had again caused intense protein decomposition, yet only 2.97 mg. of the released nitrogen had appeared in the form of amides. Of the infiltered ammonia 3.87 mg. had thus been metabolised to amides, the carbon framework of which was drawn from some nonprotein source. Schwab concluded that the succinate-ion, contrary to the idea of Mothes, need have played no essential role in the other case quoted.

His next experiment was with prophylls of *Phaseolus multiflorus* which had been kept for two days in the dark. These were infiltered with 0.05 M. nutrients buffered at pH 6.4.

TABLE 71.
(From Schwab, 304.)

Prophylls of *Phaseolus multiflorus* (fresh weight 28 g.) Infiltered with:	H_2O	After 21 hours in the dark NH_4 malate	NH_4 oxalate	NH_4 sulphate
	mg.	mg.	mg.	mg.
Ammonia-N	1.12	1.89	6.30	1.64
Amide-N	5.60	9.76	12.20	9.65
Protein-N	126.00	126.00	105.00	100.80
Increase in amide-N	–	4.16	6.60	4.05

In the case of the oxalate, very little ammonia had been used and the increase in amide nitrogen was clearly due to proteolysis. The malate and sulphate, however, show equal increases in amide nitrogen and, if the respective leaves were infiltered with equal amounts of ammonia (as I assume), then the carbon skeleton of the amide may have come in both instances from the same source.

In another experiment, he infiltered etiolated wheat seedlings with ammonium bicarbonate and also with the sodium salts of aspartic and glutamic acids, the treatment being repeated after two days in the dark.

TABLE 72.

(From Schwab, 304.)

3-week etiolated seedlings of *Triticum sativum* (fresh weight 40 g.)

After 4 days in the dark

Infiltered with:	H_2O	NH_4HCO_3	Na aspartate	Na glutamate
	mg.	mg.	mg.	mg.
Ammonia-N	0.84	7.17	2.13	1.68
Glutamine amide-N	0.42	3.12	3.36	3.00
Asparagine amide-N	9.96	13.64	13.77	13.51
Protein-N	95.20	105.00	112.00	112.00
Increase in total amide-N	–	6.08	7.45	6.83

The amounts of both amides formed in each case being the same, Schwab concluded that the aspartic acid and glutamic acid had first been deaminated and the resulting ammonia used for amide synthesis in the usual way. With this conclusion I do not entirely agree, for I think that Schwab has failed to take into account the fact that not only was protein being synthesised in the growing parts but protein decomposition was (presumably) still going on in the endosperm. It seems to me that we have here a parallel case to the experiments of Prianischnikow which I discussed in Chapter IV (p. 100): in the growing parts, synthesis of protein is taking place preferentially at the expense of amino acids and the observed increase in amide nitrogen may be due to the continued decomposition—perhaps stimulated by the infiltration—of the reserve protein in the endosperm, for Schwab himself shows (cf. his table 4) that the endosperm protein of wheat gives, on germination, glutamine as well as asparagine amide nitrogen. In the absence of data concerning the changes in amino nitrogen and in residual protein of the endosperm, I cannot profitably discuss this experiment further.

Schwab felt that his numerous experiments had failed to substantiate Mothes's ideas, and to clinch the matter, if possible, he attempted to follow the fate of the infiltered organic acid (by change of *total acidity*). The first experiment with *Lappa tomentora* and ammonium succinate showed that much more acid was used up than the equivalent of asparagine formed, and the same was true of malate, which he used because (he states) it is more resistant to the attack of dehydrases. The experience, however, showed that these leaves were not well suited for this purpose, and he made further experiments with leaves of the runner bean which had been kept in the dark for three days to deplete them of carbohydrate reserves. Using 0.1 M. ammonium malate, an approximate equivalence was found (column 1).

TABLE 73.

(From Schwab, 304.)

Phaseolus multiflorus	Per 100 g. fresh leaves	
	(1) 3-day darkened leaves infiltered and kept 24 hours in the dark	(2) 4-day darkened leaves infiltered and kept 24 hours in the dark
	mg.	mg.
Asparagine synthesised	456.1	5.4
Malic acid infiltered	597.8	298.7
Malic acid left	200.2	78.8
Malic acid used up	397.8	220.1

In a further experiment, however, with leaves infiltered this time with 0.05 M. potassium calcium malate (column 2), very little amide was formed, although a large amount of acid was consumed. Schwab admits that the latter might have been used for energy metabolism, thus sparing the oxidative breakdown of protein.

Summarising his own work, Schwab stated that he could find no evidence for the essential co-operation of C_4-dicarboxylic acids in asparagine formation, as Mothes had suggested; on the contrary, amide formation appeared to depend only on the availability of the ammonia and the carbohydrate level, the dependence of the carbon skeleton on the latter being shown by the synthesis of glutamine when leaves are infiltered with ammonium succinate or malate.

While admitting the validity of Schwab's criticisms of Mothes's conclusions, and of the deductions he draws from some of his own experiments, I cannot agree that the latter provide any evidence whatsoever as to the role of dicarboxylic acids in amide synthesis. It is to the merit of Schwab that he has recognised (again without making any *ad hoc* analyses!) that leaves kept for some time in the dark may still contain carbohydrate reserves, but he, like Mothes, has overlooked the possibility that an immediate C_4-dicarboxylic acid precursor (C_5 for glutamine?) or a similar acid from which it could be readily produced was already present in the leaves and, in his case, in amount sufficient to combine with all the available ammonia. Mothes's results quoted in tables 68 and 69 show a synthesis of about 0.5 g. and 0.6 g. of new asparagine per 100 g. of fresh leaf, respectively, amounts very much greater than those given in each of Schwab's experiments except the one example quoted in table 73, wherein, incidentally, he found an approximate equivalence with the malic acid used! Assuming that Mothes's bean leaves contained 16 per cent of total solids, then 0.6 g. of asparagine would need about 4 per cent of a C_4-dicarboxylic acid precursor (on a basis of dry weight); this might be considered high, and therefore evidence that part of the asparagine had come from the infiltered acid. On the other hand, Vickery, Pucher, Wakeman and Leavenworth (336) found 15 per cent of malic acid in leaves of *Nicotiana tabacum,* and, if this acid can function in asparagine synthesis, the small amount of the latter formed when Mothes infiltered such leaves with ammonium pyruvate and lactate needs no further comment! (Cf. his table 10, p. 137.)

In my opinion, the experiments of both Schwab and Mothes provide no definite evidence at all as to the precursors of asparagine or glutamine since, as I have stated above, the necessary analyses of the pre-formed organic acid present in the leaves have not been made. I am not denying that the plant can readily synthesise the immediate precursors—be they C_4- or C_5-dicarboxylic acids or other, as yet unidentified, products—from plant reserves such as carbohydrate; all that I insist upon is that we explore the possibilities of those plant acids *known* to stand in close chemical relationship to the amides, and to be highly reactive in other biological systems, before invoking the aid of *unknown* products.

This attitude was forced on me prior to the publication of Schwab's paper by some experiments that I had been conducting since 1934 on the synthesis of glutamine in perennial rye-grass (*Lolium perenne*).

In the spring of 1933, Greenhill observed that certain of his pot-cultures of this grass, which had received a heavy dressing of ammonium sulphate the previous day, had a strange appearance and looked as though a whitewash brush had been drawn over the ends of the blades. Closer examination revealed the presence of a white substance adhering firmly to the ends, and sufficient material was eventually collected to show that it was glutamine (Greenhill and Chibnall, 99). This gave me the suggestion that blades of rye-grass, which must be capable of elaborating glutamine with extraordinary rapidity, might be suitable for infiltration experiments on lines similar to those of

Mothes and reveal the precursor of the 5-carbon skeleton in this amide. Solutions of ammonium glutamate, di-ammonium glutarate and di-ammonium glutaconate of 0.1 M. concentration were used.

TABLE 74.

(Unpublished data.)

Blades of *Lolium perenne* (4.vii.34)
Temp. 16–17°

Per 10 g. fresh blades

Substance unfiltered	Time in hours	Total soluble-N	Ammonia-N	Glutamine amide-N	Asparagine amide-N	Amino-N (Van Slyke) after removing free ammonia
		mg.	mg.	mg.	mg.	mg.
Water	0	14.2	0.4	0.6	0.5	7.1
"	22	19.8	0.2	1.7	0.5	9.1
Ammonium glutamate	0	31.7	8.5	0.6	0.4	15.4
"	22	35.2	1.4	7.6	1.2	21.8
Ammonium glutarate	0	31.7	16.0	0.4	0.7	6.8
"	22	35.4	12.8	3.2	1.3	11.8
Ammonium glutaconate	0	30.6	14.6	0.5	0.7	6.8
"	22	35.0	11.2	4.4	1.5	13.9

In the case of the ammonium glutamate, the response was excellent, and since glutamine gives about 90 per cent of its *amide* nitrogen as amino nitrogen by Van Slyke's method, it can be shown by calculation (allowing for the control) that the new glutamine has been formed by dehydration of the ammonium glutamate, thereby showing that the blades contain an actively synthesising *glutaminase*, an enzyme whose presence in the roots of beet has been recently demonstrated by Vickery *et al.* (334) and in animal tissues by Krebs (125). The response with the other two salts, although positive, was not very great, and the next experiments, using ammonium pyruvate, would have seemed much more encouraging if, at the same time, another batch of blades had not been infiltered with di-ammonium hydrogen phosphate.

TABLE 75.

(Unpublished data.)

Blades of *Lolium perenne* (18.iii.35) Temp. 16–18°

Per 10 g. fresh blades

Substance infiltered	Time in hours	Total soluble-N mg.	Ammonia-N mg.	Glutamine amide-N mg.	Asparagine amide-N mg.	Amino-N after removing free ammonia mg.	Amino N after 3 hours at 100° at pH 6 and removing ammonia mg.
Water	0	13.2	0.2	0.9	0.5	5.9	4.7
"	22	18.4	0.2	2.3	0.6	9.7	6.2
Ammonium pyruvate	0	21.2	7.6	0.9	0.8	–	–
"	22	24.3	1.0	5.5	0.8	16.3	5.9
Ammonium phosphate	0	24.7	9.9	0.8	0.8	–	–
"	22	27.2	2.3	6.0	0.9	17.8	6.0

In both cases there was a strong synthesis of glutamine, and the presence of the amide was proved by the fall in amino nitrogen equivalent to about 90 per cent of the total glutamine-N on heating for three hours at pH 6 (Chibnall and Westall, 52). Clearly the non-nitrogenous precursor required for the glutamine either pre-existed in the blades, or was readily synthesised from reserves, and our experiments with the ammonium salts of pyruvic, glutaric and glutaconic acid had not given the required information, for there was no evidence that the acid-ions had been utilised at all!

Suspecting that the precursor was actually an organic acid, the focus of our research was, for a time, directed towards these substances, and, as the infiltration experiments cannot be conveniently carried out on samples of leaves greater than 10–20 g., we sought for a method of estimating the total organic acids in not more than 5 g. of fresh leaf material, and the one* finally adopted was a modification of that of Pucher, Vickery and Wakeman (234, 235).

We then infiltered blades of perennial rye-grass with the ammonium salt of α-ketoglutaric acid (prepared by titration to pH 6.8). This acid, which was not available in our earlier experiments, was used to test reactions 2 and 5 (pp. 106, 108) and we were able to follow its metabolism by a titration of the ether extracted acids for bisulphite-binding capacity, the method used for this purpose being that of Clift and Cook (54), as modified by Elliott, Benoy and Baker (75).

A preliminary experiment showed that leaves infiltered with the ammonium salt and kept for 20 hours in the dark showed no visible injury, while the proteolysis brought about was no greater than in the corresponding water-infiltered leaves. Moreover, the glutamine response was excellent, and the whole of the metabolised ammonia was accounted for as nitrogen of new glutamine, together with a subordinate amount of new asparagine.

In the next experiment, the 20 g. samples of fresh leaf material were rapidly dried in a current of air at 80° before analysis.†

* The dried and finely powdered leaf material (0.2 g.) is ground with (usually) 0.2 ml. 5 N hydrochloric acid and sufficient water to make the total volume of liquid 0.4 ml., until a uniform mash is obtained. Six grams of washed sand are then added and the mixture is ground to a paste, which is then transferred to a Soxhlet thimble and extracted with ether continuously for 30 hours. The ether extract is concentrated, poured into warm water—which removes the residual ether—and the aqueous solution filtered to remove lipoids. The slightly brown filtrate is then made up to standard volume, the total organic acid and mineral acid estimated by titration to pH 7.5 using α-naphtholphthalein, and the mineral acid alone by titration in the presence of 9 volumes of acetone to pH 4.0, using naphthylamine orange (cf. Richardson, 242). Phosphate and sulphate ions (normally absent) must be allowed for. I wish to thank Dr. G. M. Richardson and Dr. J. W. K. Lugg for considerable assistance in devising this method.

† The aqueous extract for analysis was prepared in the following

TABLE 76.
(Unpublished data.)

(Figures given are in mg. per 10 g. fresh blades.)

Lolium perenne
(16.vi.38) Temp. 20°

	Water			Di-ammonium α-ketoglutarate					
	(1)	(2)	(3) (2−1)	(4)	(5)	(6) (5−4)	(7)	(8) (7−4)	(9) (8−3)
Hours	0	20		0	4		20		
Ammonia-N	0.4	0.8	+0.4	8.1	5.6	−2.5	2.3	−5.8	−6.2
Total glutamine-N	0.5	2.7	+2.2	0.4	3.6	+3.2	7.8	+7.4	+5.3
Total asparagine-N	0.5	1.0	+0.5	0.5	1.0	+0.5	2.4	+1.9	+1.4
Sum of above	1.4	4.5	+3.1	9.0	10.2	+1.2	12.5	+3.5	+0.4
Amino-N (Van Slyke)*	3.4	5.4	+2.0	3.5	−	−	9.0	+5.5	+3.5
Protein-N	46.5	38.6	−7.9	45.0	−	−	34.8	−10.2	−2.3
Ketonic organic acids†	26.8	26.8	0	56.8	39.7	−17.1	30.6	−26.2	−26.2
Non-ketonic organic acids†	76.2	76.2	0	78.2	84.3	+6.1	92.4	+14.2	+14.2
Reducing sugars	36	102	+66	32	31	−1	64	+32	−34
Non-reducing sugars	478	413	−65	482	458	−24	390	−92	−27

The blades infiltered with the ammonium salt had undergone a little more intense proteolysis than those infiltered with water, but even so the data given in columns 6 and 9 (table 76) showed that the major part of the metabolised ammonia nitrogen had reappeared as glutamine nitrogen. The amino nitrogen figures confirm this. Ketonic acid had disappeared,‡ and, on the assumption that

way. 0.5 g. of dried material was ground with 50 ml. of water and 1 ml. of 5 per cent acetic acid, the mixture warmed rapidly to 80°, cooled rapidly and filtered. The residue was again treated in a similar way. No detectable amount of starch passed into solution.

* Corrected for abnormality of glutamine.
† Calculated for convenience as α-ketoglutaric acid (m.w. 146).
‡ More direct evidence that the infiltered α-ketoglutaric acid had been metabolised, and not other material normally present in the blades which exhibits bisulphite-binding capacity, was obtained from a parallel experiment in which the ketonic acids were separated by 2,4-dinitrophenylhydrazine. The extract prepared from the blades immediately after infiltration with the ammonium salt gave a voluminous precipitate from which the 2,4-dinitrophenylhydrazone of α-ketoglutaric acid (m.p.

it had been metabolised to glutamine, the agreement between the calculated and observed values is very fair.

TABLE 77.

(From columns 6 and 9, table 76.)

(Figures given are in mg. per 10 g. fresh blades.)

Hours	4	20
Decrease in α-ketoglutaric acid	17.1	26.2
Calculated equivalent increase in total glutamine-N	3.3	5.0
Observed increase in total glutamine-N	3.2	5.3

We had thus obtained convincing proof that, on infiltration of blades of perennial rye-grass with this ammonium salt, the ammonia had been utilised in the synthesis of glutamine, and to a much lesser extent in that of asparagine, while the amount of α-ketoglutaric acid which had disappeared was approximately equivalent to that which could have been utilised in the glutamine synthesis. Had we, accordingly, obtained convincing proof that the α-ketonic acid provided the carbon skeleton of the glutamine? A study of certain other data provided in table 76 shows that conditions have again been too complex to permit of such a straightforward deduction without further consideration.

All starch disappeared from the blades infiltered with water during the 20 hours in the dark (colour test); this was accompanied by an increase in reducing sugars and a decrease in non-reducing sugars (column 3, table 76), the gross loss being due to respiratory activities. Organic acids were unchanged. Starch had also disappeared from the blades infiltered with the ammonium salt, but, compared with the water-infiltered control, there was a loss in both reducing and non-reducing sugars (column 9, table 76). There was also an increase in the amount of non-

122°) was readily obtained by fractional crystallisation, whereas that made from blades treated in a similar way, but kept for 20 hours in the dark, and also from blades infiltered with water, gave a small amount of a dark-brown gum from which nothing crystalline could be obtained.

ketonic acid present. Since 61 mg. of sugar had disappeared, it might be held that the carbon skeleton of the glutamine had come from this source, and the α-ketoglutaric acid had been oxidised to succinic acid or had undergone dismutation to give α-hydroxyglutaric acid, thus accounting for the increase in non-ketonic acids. This is very unlikely for two reasons. In the first place the synthesis of glutamic acid requires the hydrogenation of α-iminoglutaric acid by dihydro-cozymase as hydrogen carrier (reaction 2, p. 106), and v. Euler *et al.* (80) have shown that the hydrogen can come into the system by dehydrogenation of glucose. In the second place, Krebs (125) has shown that in animal tissues the energy required for the synthesis of glutamine from ammonium glutamate—which is an endothermic reaction—is obtained through respiration, and that in e.g. brain and retina the rate of synthesis is slow in the absence of glucose. There are thus reasonable grounds for assuming that in the infiltered blades of perennial rye-grass a synthesis of glutamine from ammonium α-ketoglutarate has definitely taken place, the hydrogen and energy for the reactions concerned having been provided through the (increased) respiration of sugars.

I have dealt with this matter at some length because the conclusion of Schwab—that he could find no evidence for the essential co-operation of C_4-dicarboxylic acids in asparagine formation, which appeared to depend only on the carbohydrate level—seems to have strongly influenced certain of the contemporary researches which I shall discuss later. Mothes's work on the fumaric-malic-succinic equilibrium in asparagine formation clearly needs repetition under conditions in which the fate of the actual infiltered acid is followed, for a determination of change in total organic acids, which was Schwab's procedure, can give a misleading result, as the data in table 76 testify.

CHAPTER X

THE ROLE OF PROTEINS IN THE RESPIRATION OF DETACHED LEAVES

I HAVE already mentioned the limitations of much of the early work on starved leaves, and it is fortunate that in the following discussion I am able to make use of the extensive data of two recent researches in which these do not apply—that of Yemm, who used barley leaves, carbohydrate rich, which were kept in a darkened chamber for some days under conditions such that carbon dioxide production could be determined; and that of Vickery, Pucher, Wakeman and Leavenworth, who used tobacco leaves, poor in carbohydrate and rich in organic acid, which were cultured in water for some days either in continuous light or in darkness.

Before describing in detail these two researches, both of which are concerned *inter alia* with the possible role of proteins in the respiration of starved leaves, it is necessary to mention briefly some early work of Deleano (63), who measured the respiration of detached vine leaves (*Vitis vinifera*) floating on water in the dark over a period of 493 hours, companion analyses being made at intervals for mono- and di-saccharides, starch, hemicelluloses, "organic acids," protein nitrogen, soluble nitrogen and ammonia. Certain of the chemical methods he employed would now be considered unsatisfactory, but his experiments undoubtedly demonstrated that vine leaves, when starved in a darkened chamber containing air saturated with water vapour, exhibit no breakdown of protein until after about 100 hours. Starch diminished during this period, but the concentration of monosaccharides remained substantially unchanged. During more prolonged periods in the dark, there was a parallel disappearance of protein and monosaccharides, and, since the carbon dioxide output was in excess of that which could be ascribed to carbohydrates, he considered that this was due to the utilisation of protein and "organic acids," which were drawn into respiration because carbohydrates were no longer available in sufficient amount for the purpose. This result was recognised, at the time, to be of importance, for it was in keeping with the idea then current that the leaf proteins were stable products which did not enter readily into metabolic changes, and it has also been frequently quoted since as evidence that breakdown of protein in leaves occurs only under conditions of carbohydrate starvation, a general conclusion which is certainly not in keeping with present-day observation.

Yemm carried out two main series of experiments: in the first (364), the carbohydrate and in the second (365), the nitrogen changes in detached and darkened Plumage Archer barley leaves were correlated with carbon dioxide production. In the former case, I quote chiefly from his experiment V, in which the samples consisted of the third leaf inserted on the axis, counting the rolled apical one as the first, from plants on which young ears were just emerging.

Measurement of carbon dioxide production did not start until 6 hours after detachment of the leaf from the plant, but determinations were made at 6-hour periods up to the 54th hour and subsequently at 12-hour intervals. The curve given in figure 4 is very close to the mean of the various samples examined, over 70 in all. It rises sharply up to the 12th hour, then falls rapidly to about the 40th hour and subsequently remains fairly steady as the leaves became yellow. Concurrent with the final browning of the leaves there was exudation of water and disorganisation of the protoplast, indicating that this phase may be considered as the "death" of the leaf cells.

Sucrose was the chief source of hexose in the leaf, as it initially formed 60–70 per cent of the total carbohydrate present. On starvation a rapid fall in the concentration of this sugar occurred; concurrently there was a complete disappearance of fructose, while the concentration of glucose rose at first, and later fell away to a low value. This

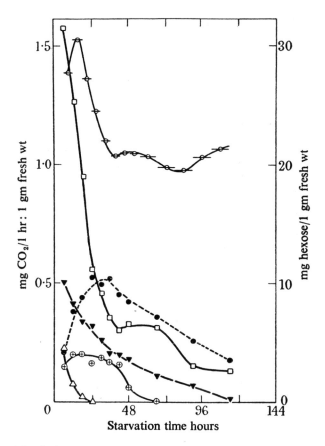

Fig. 4. *Carbohydrate exhaustion and carbon dioxide production.* △, *fructose;* ⊕, *fructosan;* ⊙, *mean carbon dioxide;* ●, *glucose;* □, *sucrose;* ▶, *starch (from Yemm, 364).*

general interrelationship of these three sugars was observed in every experiment. Starch disappeared slowly and continuously, so that at the end of the experiment,

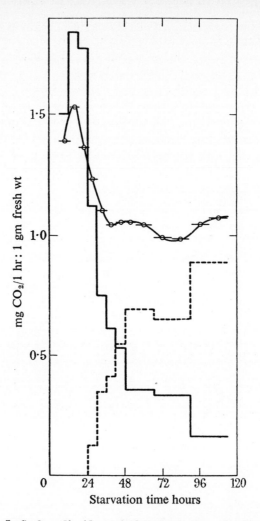

Fig. 5. Carbon dioxide equivalent to the loss of total carbohydrate (Experiment V). ⊙, mean CO_2 output; solid line, carbon dioxide equivalent to carbohydrate loss, i.e., the rate of carbohydrate loss calculated as mg. carbon dioxide per hour per gm. of fresh weight; dotted line, carbon dioxide not accounted for by carbohydrate loss (from Yemm, 364).

when the leaves were 75 per cent yellow, it could no longer be detected.

The rate of total sugar loss was calculated from the difference between successive analyses, and it is clear that only in the first stages can the carbon dioxide be considered to arise solely from these sources. Progressively with starvation time, the respiration was taking place at the expense of undefined sources (indicated by the dotted line in fig. 5), and throughout the yellowing period, from

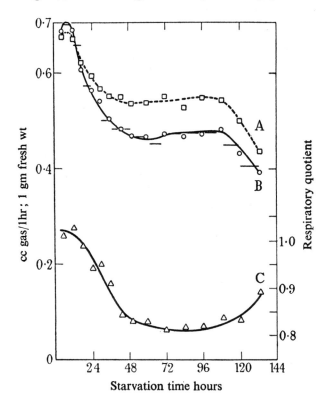

Fig. 6. *Carbon dioxide production and oxygen uptake during starvation. Carbon dioxide (Pettenkofer method),* — *carbon dioxide (Haldane method),* ○ ; *oxygen,* □. *Curve A, oxygen uptake; curve B, carbon dioxide output; curve C, respiratory quotient (from Yemm, 364).*

the 40th to the 140th hour, less than 25 per cent can be accounted for by carbohydrate loss. Yemm remarks that in certain other experiments he found that, even during the first 12 hours, the estimated hexose loss did not account for all of the carbon dioxide produced, the utilisation of other products depending, primarily, on the initial level of available carbohydrate, being delayed if this were high.

Confirmation that carbon dioxide was being actively produced from undetermined sources was obtained by measurement of the respiratory quotient. Up to the 20th hour after detachment of the leaves, a quotient very close to unity was recorded, and from the 20th to the 40th hour a transition occurred to a value of about 0.8. The lower value was maintained during the yellowing phase, but a rise was observed after 120 hours concurrent with the final rapid fall in rate of carbon dioxide production. This rise in the respiratory quotient must be associated with the rapid browning of the leaf, and the condition of the majority of the cells at this phase indicated that the connection between these oxidations and normal cell respiration must have been remote. Suspecting that the unknown source of carbon dioxide production might be protein, Yemm repeated these experiments the following year, attention being paid to nitrogen changes. In this case both Plumage and Spratt Archer varieties were used, grown under slightly different experimental conditions, but the results of carbon dioxide production and nitrogen analyses indicated that neither varietal nor cultural differences were serious enough to make a direct comparison difficult.

The rate of protein breakdown, as indicated by decrease of insoluble nitrogen, seemed to be greatest during the early stages of starvation and no true sparing action of carbohydrate was observed, since during this initial phase the carbohydrate concentration was at its greatest. Yemm suggests that a continuous hydrolysis of protein may be a normal mechanism whereby proteins, synthesised in the mature leaf, are translocated to other parts of the plant—citing previous speculations of my own (36, 37) which I shall discuss more fully later.

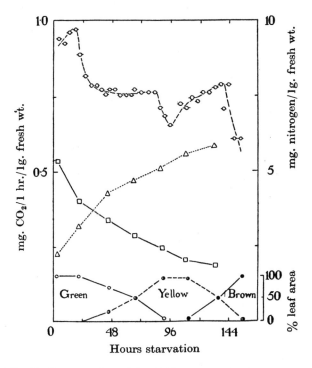

Fig. 7. *Changes of total insoluble and total soluble nitrogen during starvation (Experiment X) and mean rate of carbon dioxide production. Yellowing and browning of the leaves shown for comparison. CO_2 production ○; total insoluble nitrogen □; total soluble nitrogen △; % area, yellow ◐; % area, brown ● (from Yemm, 365).*

The decomposition of protein was accompanied by a corresponding increase in the soluble nitrogenous products and in all his experiments Yemm found that during approximately the first 24 hours the only considerable change involved the amino and glutamine amide fractions. After about 24 hours, asparagine amide nitrogen increased and rose to a high value during yellowing, while free ammonia accumulated slowly after about 48 hours. A final phase was characterised by a rapid formation of

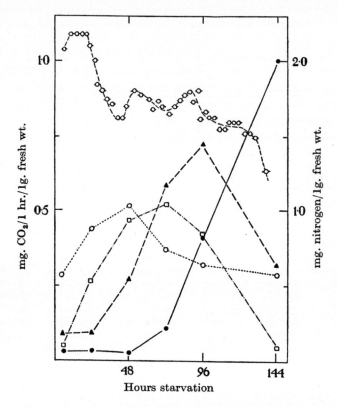

Fig. 8. Changes in amino, amide and ammonia fractions during starvation (Experiment XI), and mean rate of carbon dioxide production. Carbon dioxide production, ⊖; amino nitrogen fraction, ○; 2 × unstable amide (glutamine) nitrogen, □; 2 × stable amide (asparagine) nitrogen, ▲; ammonia nitrogen, ● (from Yemm, 365).

ammonia, considerable quantities escaping as gas after browning of the leaf and significant losses probably occurred from the onset of browning, after about the 120th hour of starvation.

In light of the general evidence available at the time, Yemm discussed the possibility that the rapid production of glutamine during the first 24 hours of starvation might be due either to primary protein breakdown, or to secondary oxidative processes involving the synthesis of glutamic acid from pyruvic acid *via* α,α'-diketoadipic acid and α-ketoglutaric acid according to the scheme of Weil-Malherbe (356) and could come to no definite decision. Dr. Yemm has since carried out another series of experiments on similar lines, and I have to thank him for permission to incorporate the following unpublished data, which show that both alternatives are possible.

TABLE 78.

(From Yemm, 365, and unpublished data.)

(Figures given are in mg. per g. fresh tissue.)

Blades of barley
Experiment IX of 1935

Period of starvation (hours)	Increase in total glutamine-N	Increase in total asparagine-N	Increase in amino-N less that due to asparagine and glutamine	Decrease in protein-N	Amide-N of decomposed protein* $\times$ 2 (calculated)
24	0.42	0.02	0.32	0.94	0.09
48	0.84	0.36	0.46	2.34	0.23
72	0.95	0.98	0.19	2.60	0.26

These results show that within the first 24 hours the amount of glutamine formed is about four times that which could have arisen by primary protein decomposition, assuming that the protein amide nitrogen be exclusively assigned to this amide. Dr. Yemm informs me that two other experiments of his 1934–1935 series gave results that were equally as emphatic.

In another case younger plants had been used and the breakdown of protein was much less rapid. It will be seen

* The amide-N of the isolated barley leaf protein is 4.8% of the total-N. (Private communication from Dr. Yemm.)

that the glutamine produced could have been of primary protein origin, again with the reservation mentioned above.

TABLE 79.

(From Yemm, unpublished data.)

(Figures given are in mg. per g. fresh tissue.)

Blades of barley
Experiment VIII of 1937

Period of starvation (hours)	Increase in glutamine amide-N	Decrease in protein amide-N*
24	0.060	0.060
48	0.125	0.115
72	0.250	0.250

These experiments of Yemm are of great interest, for they show that breakdown of protein is discernible within 24 hours of the leaf being detached from the plant, and that, under certain conditions, the amino acids resulting from this decomposition have by then already undergone oxidation to give ammonia for glutamine production. As Yemm points out, the effect of such an oxidation on the total gaseous exchange would be scarcely beyond the limit of accuracy with which the respiratory quotient was determined.

In accordance with his deductions from carbohydrate analyses and measurements of the respiratory quotient, Yemm found that the time at which asparagine began to accumulate corresponded closely with that at which the breakdown of protein appeared to be important in carbon dioxide production. Such a transformation, nevertheless, should give a respiratory quotient of only 0.7 as against an observed value of over 0.8 and he discussed at some length the possible reasons for the discrepancy. The interpretation of his results was, however, incomplete, for he had not (at the time) investigated the fate of the organic

* Determined by hydrolysis of the leaf residues with 5% HCl for 3 hours. The amide-N was 5.1% of the total-N. (Private communication from Dr. Yemm.)

acids present in the barley leaves. It is for this reason that the pioneer work of Vickery, Pucher, Wakeman and Leavenworth (336) is so valuable, for it provided us, for the first time, with a complete picture of the organic acid metabolism in the starving leaf.

These latter workers used leaves of Connecticut tobacco grown under shade, which were chosen at random from the three lowest fully formed leaves growing on the plants. They carried out six separate experiments, in which the leaves were placed with their cut ends in water, glucose solution and an ammonium salt nutrient solution, respectively, either in continuous darkness or in continuous light (artificial light was supplied at night). Each sample consisted of 60 leaves, the initial fresh weight being 1500–1660 g., so that sampling errors alone were appreciable and one is constrained to overlook small variations in composition.

At the end of 73 hours, those cultured in water in the dark were yellow along the margins and veins and the yellowing became more extensive with lapse of time, so that, at the end of the experiment (143 hours), the whole blade was yellow and the tips had already started to brown. In the corresponding case of those cultured in the light, however, only a few of the leaves had yellowed at the tips and margins after 143 hours. As was to be expected, the changes in the chemical composition of the two sets of leaves were different. Figures 9–12 refer to the samples cultured on water in the light (LW) and in the dark (DW).

The leaves had been picked from the plants fairly early in the morning, when the content of sugars and starch was very low. Those cultured in the light showed rapid photosynthesis, the organic matter (per 1000 g. fresh weight) increasing from 74.6 g. to 102.4 g. in 143 hours. This was due in part to increases in total (both fermentable and unfermentable) sugars and in starch, but other, undetermined soluble products were also formed. In this particular experiment glucose and sucrose were not separately determined, but Dr. Vickery has kindly supplied me with data from a similar, but later, experiment, which show that the main increase in fermentable sugar is due to glucose (fig. 12). Here the increase in total organic solids

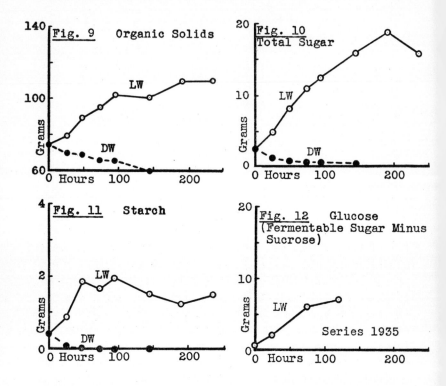

was satisfactorily accounted for in terms of total sugars and starch. From these results there can be no doubt that the leaves had been collected at a stage in the development of the plants when photosynthesis was very active, and the products were being almost completely translocated away from them at night. In contrast to those cultured in the light, those cultured in the dark showed a substantial fall in total organic solids (from 72.5 g. to 56.15 g. per 1000 g. fresh weight) and both starch and sugar rapidly disappeared.

TABLE 80.

(From Vickery, Pucher and colleagues, 336.)

Leaves of *Nicotiana* *Tabacum* Hours	Unless stated, in g. per 1000 g. fresh leaf					
	0	25	49	73	95	143
Cultured on water in light						
Malic acid	15.20	15.00	15.80	16.20	15.10	16.40
Citric acid	3.26	3.25	3.24	4.02	3.68	3.50
Oxalic acid	1.57	1.62	1.64	1.57	1.65	1.65
Undetermined (mg. equivalents)	−3.00	−0.20	21.00	15.70	28.40	34.80
Cultured on water in dark						
Malic acid	15.20	12.00	10.70	6.55	6.02	4.48
Citric acid	3.26	4.88	5.20	7.08	8.73	9.45
Oxalic acid	1.57	1.56	1.61	1.76	1.84	1.64
Undetermined (mg. equivalents)	−3.00	16.10	47.50	58.40	29.90	28.20

The organic acids which were determined by the methods of Pucher and his colleagues (233, 235, 236) likewise exhibited an interesting difference in the two cases. Vickery, Pucher and their colleagues remark that, according to their experience, the organic acid content of leaves varies with age, and, as the leaves used in the different experiments cited were not all of the same age, minor differences may not be significant. In spite of this, however, many interesting points emerge. The total organic acid content, for instance, was enormously high—20.1 g. per 1000 g. of fresh leaves, equal to 22 per cent of the total solids and 27 per cent of the total organic matter present. During culture in the light, there was a small, and they think a significant, increase in organic acids of unknown constitution (the total initial organic acidity was 311 mg. equivalents), but the amounts of malic, citric and oxalic acids were unchanged. In tobacco leaves, therefore, they thought it probable that, in the presence of sufficient carbohydrate, organic acids, in spite of the large amounts present, are not directly utilised in respiration, and respiration, indeed, may even lead to an increase in acids other than malic acid, citric acid and oxalic acid.

In the dark, on the contrary, there was no significant change in total acidity, but marked changes occurred in all the components and, as I shall show later, organic acids had undoubtedly been drawn into metabolism.

The changes in the nitrogenous constituents were also of great interest, the variation in the total nitrogen of each series illustrating the magnitude of sampling errors when comparing leaf material of this type.

The digestion of protein with the formation of amino acids proceeded rapidly and at approximately the same rate for the first 73 hours in both cases; later, the rate diminished in the leaves kept in the light, but was maintained in those kept in the dark. The water-soluble nitrogenous products reflected these changes, but the resulting amide metabolisms differed. Asparagine formation was prompt and rapid in the dark, but was retarded and was much less extensive in the light; glutamine formation, on the other hand, was rapid in the light and took place scarcely at all in the dark. Ammonia formation was of importance only in the later stages of leaves kept in the dark.

Vickery (333), in a later review of this work, has shown, by comparison of appropriate data, that amide formation took place through the oxidation of amino acids to give ammonia, which then combined with the necessary precursors to produce asparagine or glutamine. In table 81, I have followed a slightly different procedure, and have compared the extent of asparagine nitrogen production with the total nitrogen set free on protein decomposition, as it seems to me that the non-amino nitrogen, as well as the amino nitrogen, will be available for the former purpose.

Vickery, Pucher and their colleagues (336) discuss at some length the possibility that the organic acids present in the leaves can act as precursors of the carbon skeletons of the amides. During culture in the light the content of

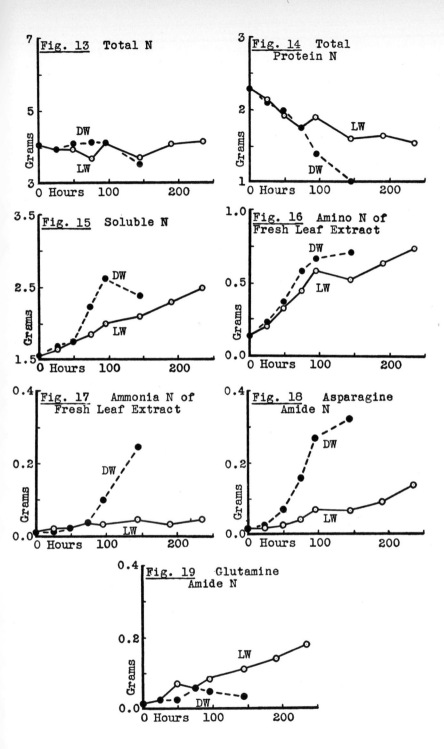

TABLE 81.

(Computed from data of Vickery, Pucher and colleagues, 336.)

Leaves of tobacco cultured on water in the dark

In g. per 1000 g. of fresh leaves

Hours	Soluble amino-N (corrected for abnormality of glutamine present) (1)	N of digested protein (2)	Δ (2 − 1)	Total asparagine-N	Total asparagine-N + free ammonia-N
25	0.066	0.180	0.114	0.012	0.015
49	0.163	0.290	0.127	0.100	0.112
73	0.227	0.560	0.333	0.278	0.308
95	0.209	0.870	0.661	0.500	0.589
143	0.221	1.270	1.049	0.604	0.838

malic, citric and oxalic acids remained unchanged and they considered that these acids could not have been utilised for this purpose; moreover, glutamine was formed when, and only when, products of photosynthesis were available, suggesting that carbohydrate or some other—unknown—substance was functioning in this case.

The origin of the carbon chain in the asparagine, however, was more obscure. In the light, this amide was formed in the presence of abundant glucose; yet it was also formed as readily in the dark when all available carbohydrate had been metabolised. In the latter case much malic acid had disappeared, and they considered the possibility that this acid might have been utilised for the purpose.

It was clear that asparagine formation could account for only a part of the malic acid loss, and, since citric acid was also formed concurrently with the asparagine, they postulated that this acid might arise from malic acid through the condensation of oxalacetic acid (reaction 8, p. 190) with pyruvic acid (formed by decarboxylation of oxalacetic acid) according to the original scheme of Mar-

TABLE 82.

(From Vickery, Pucher and colleagues, 336.)

In g. per 1000 g. fresh leaves

Tobacco leaves cultured in the dark for 143 hours on	Pro- tein-N loss	Aspara- gine increase	Malic acid equiva- lent	Actual malic acid loss
Water	1.27	2.85	2.89	10.66
Glucose solution	1.15	2.69	2.73	10.60
Ammonium salt nutrient so- lution	1.16	3.61	3.67	6.50

tius and Knoop (reaction 10, p. 191). Two molecules of malic acid would thus yield one molecule of citric acid, one molecule of water and two molecules of carbon dioxide, one molecule of oxygen being taken up. Table 83 shows the changes which would occur if such a transformation were operative.

TABLE 83.

(From Vickery, Pucher and colleagues, 336.)

In g. per 1000 g. of fresh leaves

Tobacco leaves cul- tured in the dark for 143 hours on	Citric acid gain	Malic acid loss	Ratio citric: malic acids	Citric acid gain calcu- lated	Yield in per cent
Water	6.20	10.66	0.582	7.63	81.3
Glucose solution	5.82	9.92	0.586	7.10	81.9
Ammonium salt nutri- ent solution	3.92	6.54	0.598	4.68	83.4

Certain inferences can be drawn from these results. It is obvious that malic acid loss and citric acid gain were connected in some way, directly or indirectly, for there was no significant loss of malic acid in the light and, cor- respondingly, no gain of citric acid. Furthermore, after 143 hours in the dark, the ratio of citric acid gain to malic acid loss was the same in all three experiments, in spite of differences in the amounts of acid concerned. Since, moreover, the actual amount of asparagine formed was very different in the third case (cf. table 82), Vickery and his colleagues query the view that malic acid, although present in such large amounts, could function as a precursor of asparagine in tobacco leaves.

These are weighty conclusions drawn from experiments carried out with far greater analytical detail than has hitherto been attempted and they deserve most serious consideration. The criticism that I offer is that these workers, in interpreting their results over one period of 143 hours (with the object, presumably, of eliminating the effect of sampling errors), have assumed that *in the dark* the metabolic processes proceed uniformly throughout. If a malic acid–citric acid transformation of the type they suggest was operative, and malic acid was not being utilised for other metabolic purposes such as amide production, then the ratio—malic acid loss to citric acid gain—should have been approximately constant throughout. This was not, however, the case if the data be considered in greater detail; as is shown in table 84, the variations are out of all proportion to sampling errors incidental to the experiments.

TABLE 84.

(Compiled from Vickery, Pucher and colleagues, 336.)

	Ratio of citric acid gain to malic acid loss		
Hours	73	95	143
Tobacco leaves cultured in the dark on			
Water	0.471	0.634	0.582
Glucose solution	0.632	0.458	0.586
Ammonium salt nutrient solution	0.283	0.458	0.598

It is true that the data of Vickery and his colleagues show that, throughout the whole period of 143 hours in the dark, protein breakdown has gone on fairly uniformly, yet I feel that, since we are forced to consider the metabolism of the whole leaf as one unit, two distinct phases might be recognised. During the first 73 hours, the leaves remained green, and the protein breakdown, except perhaps in the last few hours, ran parallel to that of corresponding samples of leaves cultured in the light. During the last 70 hours, however, the leaves were yellowing rapidly but not uniformly, and even though, as Yemm's results show, respiration may have remained active during this period, complete chloroplast disintegration must have been taking place in an increasing number of the leaf cells, leading, possibly, to more deep-seated metabolic changes than had occurred during the earlier period.

The organic acid changes dissected in this way bring out some interesting facts.

TABLE 85.

(From data of Vickery, Pucher and colleagues, 336.)

In g. per 1000 g. of fresh leaves

Tobacco leaves cultured in the dark on	Citric acid gain	Malic acid equivalent (2 mols.)	Excess malic acid available	Asparagine increase	Malic acid equivalent
(First 73 hours)					
Water	3.82	5.40	3.25	1.33	1.35
Glucose solution	3.68	5.14	1.00	1.27	1.29
Ammonium salt nutrient solution	1.27	1.77	2.86	1.85	1.88
(Last 70 hours)					
Water	2.37	3.31	−1.24	1.55	1.57
Glucose solution	3.08	4.30	−0.49	1.83	1.88
Ammonium salt nutrient solution	2.64	3.69	−1.78	2.38	1.42

During the first 73 hours there was, with one possible exception, ample malic acid to provide for the conversion to citric acid (2 mols. malic acid → 1 mol. citric acid) and also for asparagine formation; during the latter period, however, both citric acid and asparagine must necessarily have arisen, in part, from other sources. Before giving my own interpretation of these results in terms of the general scheme set out on page 194 (reaction 13), it is necessary to mention certain ancillary deductions I have drawn from the very extensive analytical data connected with these tobacco experiments.

In the first place it seemed to me worth while to explore the possibility that some further insight into the relationship between protein decomposition, amide formation and respiration might be gleaned from a study of the carbon as well as of the nitrogen changes which took place in the leaves cultured on water in the dark. Dr. Vickery very kindly placed samples of the necessary dried leaf material at my disposal for this purpose, and these were accordingly submitted to macro-combustion analysis for carbon, hydrogen and ash.* The results are shown in table 86.

TABLE 86.

(Unpublished data from material supplied by Dr. Vickery.)

Tobacco leaves cultured on water in the dark — In g. per 1000 g. fresh leaves

Hours	0	25	49	73	95	143
Total carbon	33.80	33.10	32.00	31.00	29.80	26.40
Total hydrogen	4.53	4.35	4.19	4.13	4.08	3.66
Total ash	19.50	–	–	20.20	22.00	20.70
Total organic solids	72.50	–	–	66.20	64.30	56.15

The changes in the various organic solids for the periods 0–73 and 73–143 hours, respectively, are collected in table 87. Here certain assumptions have been made in cal-

* Vickery, Pucher and their colleagues (336) call attention not only to the magnitude of sampling errors when working with leaves of this type, but also to the difficulty of determining the ash content of the leaf samples. I mention this because the percentage of ash found on combustion in oxygen differs slightly from that recorded by these workers. In each case the ashes contained a small amount of carbonate, due possibly to the organic acids originally present in the dried leaves. The error thus introduced in the determination of true ash, carbon and total organic solids is assumed to be constant in the series and has not been allowed for.

culating the weight of some of the nitrogenous fractions; nitrate and ammonia nitrogen are omitted and the residual organic nitrogen is regarded as mostly combined in basic substances the weight of which can be estimated as nitrogen $\times$ 5. The "undetermined" fraction includes the "unknown" organic acids mentioned in table 80 and is, of course, subject to the experimental errors of all the individual analyses and computations, and the values quoted clearly have no numerical significance.

TABLE 87.

(Computed from Vickery, Pucher and colleagues, 336, and additional data.)

Tobacco leaves cultured on water in the dark — In g. per 1000 g. of fresh leaves

	(1)	(2)	(3)	Δ (2 − 1)	Δ (3 − 2)
Hours	0	73	143		
Starch	0.43	–	–	−0.43	–
Protein (N × 6.25)	14.19	10.94	6.25	−3.25	−4.69
Insoluble organic material other than protein and starch	29.88	29.36	26.80	−0.52	−2.56
Total sugar	2.51	0.67	0.50	−1.84	−0.17
Citric acid	3.26	7.08	9.45	+3.82	+2.37
Malic acid	15.20	6.55	4.48	−8.65	−2.07
Oxalic acid	1.57	1.76	1.64	+0.19	−0.12
Asparagine	0.18	1.50	3.05	+1.32	+1.55
Glutamine	0.12	0.60	0.39	+0.48	−0.21
Nicotine	1.04	0.97	0.87	−0.05	−0.10
Amino acids (N × 7)	0.67	2.26	2.22	+1.59	−0.02
Residual organic nitrogenous compounds (N × 5)	2.75	2.30	1.80	−0.45	−0.50
"Undetermined"	0.70	2.21	−1.30	+1.52	−3.52
Total organic solids	72.50	66.20	56.15	−6.30	−10.05

The data of Vickery, Pucher and their colleagues set out in this way provide us with the most complete picture available, not only of the soluble products present in a leaf, but of the changes which these undergo during starvation in the dark—a great tribute to the fine analytical skill displayed. In drawing up the following balance

232 PROTEIN METABOLISM IN THE PLANT

sheets showing the changes in carbon during the two periods concerned, I have assumed (i) that the leaf proteins contain 50 per cent carbon and that, on digestion to give amino acids, 18 per cent of water is added; (ii) that the amino acids (average carbon = 42 per cent) resulting from protein digestion that are no longer present as such have been deaminised and the nitrogen-free amino acids thus produced are available for further metabolism; (iii) that the small amount of glutamine produced is of primary protein origin, and (iv) that the loss in "insoluble material other than protein and starch" during the period 73 hours to 143 hours, when the leaves were yellowing, is due to the breakdown of glycerides and phosphatides (carbon = 65 per cent). This is in keeping with the results of my own researches, using detached leaves cultured on water in the dark (113, 51).

Carbon balance sheet for the period 0–73 hours.

Available for metabolism	Carbon g.	To be accounted for	Carbon g.
Carbohydrates	1.12	Respiration	2.80
Malic acid	3.10	Asparagine	0.50
Excess amino acids derived from protein	0.95	Citric acid	1.46
Residual bases	0.18	"Undetermined"	0.59
	5.35		5.35

Let us consider the changes which have occurred during this period in terms of the general scheme given on page 194 (reaction 13). Citric acid has accumulated and malic acid has disappeared: if, therefore, the Krebs and Johnson cycle has been operative, the reaction rate of citric acid synthesis from malic acid (steps c and a) has exceeded that of citric acid decomposition (step b). The actual synthesis of the citric acid itself (reaction 10, p. 191) may have taken place in two different ways:

(i) At the expense of two molecules of malic acid, as

Vickery, Pucher and their colleagues have suggested. The data given in table 85 show that ample malic acid was available for such a purpose, and if it be assumed that this acid has also provided the carbon skeleton for asparagine, there would still have been a surplus of 1.9 g. which must have been utilised in other processes. The carbon balance sheet shows that this surplus, as also all the carbohydrate, would have been expended in respiration.

(ii) At the expense of one molecule of malic acid (for oxalacetic acid), the pyruvic acid or some other equivalent product being provided by carbohydrate, or by protein *via* amino acid, because secondary asparagine formation provides evidence that nitrogen-free amino acid residues have been made available for use either in step e or step d. Only 0.97 g. of malic acid carbon would have been required for the citric acid synthesis in this way, and if carbohydrate had been used in respiration through the operation of the cycle, then asparagine formation must have depended on either nitrogen-free amino acid residues or the excess malic acid. Even so, the major part of the latter must have been used in respiration.

Carbon balance sheet for the period 73–143 hours.

Available for metabolism		To be accounted for	
	Carbon		Carbon
	g.		g.
Glycerides and phosphatides	1.80	Respiration	4.60
Amino acids	2.53	Citric acid	0.87
Residual bases	0.20	Asparagine	0.56
Malic acid	0.70		6.03
"Undetermined"	0.80		
	6.03		

Turning to the second period of starvation, we see from the carbon balance sheet that amino acid residues must have been in large part utilised in respiration, thereby providing definite proof for the deductions of Deleano and Yemm that on carbohydrate exhaustion protein can be respired. As before, citric acid has accumulated, and the data already given in table 85 have shown that insufficient malic acid was available to permit an exclusive synthesis from this source. There was, however, sufficient to provide for the oxalacetic acid alone and the pyruvic acid, or its equivalent, might have been derived from other products.

TABLE 88.

(From data of Vickery, Pucher and colleagues, 336.)

Tobacco leaves cultured in the dark (last 70 hours) on	In g. per 1000 g. of fresh leaves		
	Citric acid gain	Malic acid equivalent (1 mol.)	Excess malic acid available
Water	2.37	1.65	0.42
Glucose solution	3.08	2.30	1.51
Ammonium salt nutrient solution	2.64	1.84	0.07

Fats and phosphatides have been metabolised; these can give succinic acid (Verkade, 329), which would enter the cycle at step d or could no doubt give fragments that would eventually enter at step e. Amino acid residues have also been utilised, and these again might either have entered the cycle at step e or, in certain cases, have given succinic acid (see pp. 95–97) and entered at step d. The asparagine carbon skeleton (oxalacetic acid or fumaric acid, steps f and g) might have come fairly directly from the latter sources, but if the validity of a cycle of the type we are now discussing be admitted, then all the products entering it—whether they be derived from protein, fat or carbohydrate—will lose all previous generic relationships, so that if, and when, the carbon skeleton for asparagine is required, it will be withdrawn from the cycle—as oxalacetic acid at step f, or as fumaric acid at step g—irrespective of any origin. The same applies to the synthesis of glutamine at step h. Such a deduction would provide us at once with a complete answer to the old question of whether proteins or carbohydrates furnish the carbon skeleton of asparagine and glutamine, for it is clear that both the former products provide no more than material for the organic acid cycle and it is from the cycle itself that certain of its particular components, which function as immediate precursors of the amides, are withdrawn. My previous discussions, on pages 95 *et seq.*, would thus demonstrate nothing beyond the fact that certain amino acids can be metabolised to another particular component of the cycle—succinic acid!

After this rather searching analysis of the data of Vickery, Pucher, Wakeman and Leavenworth for the tobacco leaves cultured on water in the dark, let me now give a more general account of the metabolic processes which I consider might have taken place in these and certain other cases.

It is generally considered that leaves normally obtain the energy necessary to maintain the stability of living protoplasm (presumably energy-requiring synthetic reactions are concerned) by the metabolism of sugars or, in certain rarer cases, organic acids. The outward expression of this metabolism is the emission of carbon dioxide, and, following the previous discussion, I shall assume that the energy is normally obtained through operation of the Krebs-Johnson cycle, which will set free carbon dioxide at steps a, b, and k.

The tobacco leaves under review, having been detached from the plant fairly early in the morning, contained very little starch or sugar and since they were at once transferred to the dark no further opportunity for photosynthesis was provided. During the first few hours in darkness these two products in large part disappeared, and they presumably passed through the cycle to provide most of the energy required during this period and were responsible for most of the carbon dioxide emitted. The operation of the cycle, which, in its simplest form, should leave the respective levels of each component organic acid unchanged, has, however, affected both the malic and citric acids; the latter has accumulated, while the former has diminished in amount such that it might have been used to provide both citric acid and energy, with the ultimate production of carbon dioxide. The reason why, in these leaves, the cycle did not operate uniformly is a subject for future enquiry. Proteins, for reasons that are considered in greater detail later, have already undergone decomposition and the resulting amino acids are being oxidised, the ammonia set free condensing with either oxalacetic acid or fumaric acid, withdrawn from the cycle, to produce asparagine, while the ketonic acids had become available for use in the cycle. In a very short time, therefore, certainly within 20 hours, all three of what we can regard as the energy-providing reserve products *viz.* carbohydrates, organic acids and proteins, have been concerned, in varying degree, in processes which have provided energy and led to the emission of carbon dioxide. Carbohydrates have undoubtedly contributed the larger share, and we can regard this as the "carbohydrate phase."

During the second period in darkness, between perhaps the 20th and 70th hours, carbohydrates are playing a minor role, and energy is now being obtained from malic acid—which is in part trapped as citric acid,—while the products of the continued protein decomposition are being almost completely oxidised (see table 81); the ammonia condensing, as before, to give asparagine and the ketonic acids being used in the cycle. We can regard this as the "organic acid phase."

In the final period, from the 70th hour onwards, the supply of organic acids has become diminished, and energy is now being obtained, to an increasing extent, from protein. The leaves are yellowing, chloroplast disintegration and general breakdown of cytoplasm are increasing and small amounts (relatively) of fats and phosphatides become available for metabolism. But the brunt of the burden is now being carried by the protein—it is the "protein phase"—and a stage is at last reached when the amino acid residues are no longer sufficient for both energy reactions and the removal of ammonia as asparagine (Prianischnikow's detoxication), and thus probably the resulting accumulation of ammonia is the chief contributory cause of cytolysis and death. The three phases distinguished above, which would be characterised by respiratory quotients of 1, > 1 and < 1, would normally overlap, and the observed quotient would not be far away from unity unless one or other of the two latter phases were very sharply defined.

The leaves cultured on water in continuous light present interesting differences. In the first place the malic acid, citric acid and undetermined acid levels show but little change throughout the experiment (235 hours), so that the cycle, if operative, has functioned normally. Photosynthesis was active from the start, and energy has undoubtedly been obtained in very large part from carbohydrate. Proteins, however, for reasons that are not immediately apparent, have been drawn into the cycle as readily as in the leaves kept in the dark, and amino acid oxidation, although initiated, perhaps, more slowly than in the latter leaves, has undoubtedly occurred, even though abundant carbohydrate was present. Possibly reflecting the greater rate at which step b was achieved, the ammonia has condensed this time with α-ketoglutaric acid to give glutamine, for asparagine production was unimportant until after 100 hours. Protein decomposition was continuous, and although the rate was depressed somewhat after about 70 hours, there was, at the same time, a parallel breakdown of chlorophyll (in keeping with many other observations mentioned elsewhere in the text), and, provided that the leaves could have been kept sterile, I think that disintegration of the chloroplasts and the resulting absence of photosynthetic products would have been the ultimate cause of cell death, as in the case of leaves kept in the dark.

In a later bulletin (338), these workers give an account of the changes taking place in detached and leaf-denuded stalks of tobacco when these were cultured on water in darkness and in continuous light. The individual sampling errors were large, and it is necessary to consider the changes over a period of two to three hundred hours. Protein decomposition occurred, and there was an increase in both amino nitrogen and glutamine nitrogen. The stalks were originally rich in fermentable carbohydrates, and the considerable loss on starvation took place at the expense of glucose (fermentable sugar — sucrose), the sucrose remaining unchanged. The citric acid and oxalic acid contents were small, and unchanged, but malic acid showed a definite increase, and, as Vickery, Pucher, Wakeman and Leavenworth point out, this must have come from the metabolism of carbohydrate.

It would appear that, in this case, we are dealing throughout with a "carbohydrate phase" and if the cycle has been operative there can have been no retardation at step b and, as in the case of the leaves cultured in the light, the ammonia resulting from the amino acid oxidation has been used in glutamine synthesis. The increase in malic acid may have come *via* step c without actual entry into the cycle.

Yemm and Somers have recently concluded an investigation of the organic acids in detached barley leaves by the methods of Pucher and his colleagues (233, 235, 236), and I have to thank Dr. Yemm for permission to make use of the (unpublished) data given in figure 20.

Malic acid and citric acid account for only about one half of the total titratable (ether-extracted) acid in the initial stages, and, since the amount of oxalic acid present was too small for positive determination, the remaining, unknown, acid, in the absence of a molecular weight, can be given no actual magnitude. Yemm's previous work (pp. 212 *et seq*.) showed that detached barley leaves, during the first 12 hours or so, were consuming much carbohydrate and the respiratory quotient was unity. It appeared to be a typical "carbohydrate phase," and, in conformity with both the cases cited above, it was accompanied by a breakdown (small, relative to the sugar consumed) of protein and (in certain cases) oxidation of the resulting amino acids. The organic acid analyses show that malic and citric acids have undergone small, but

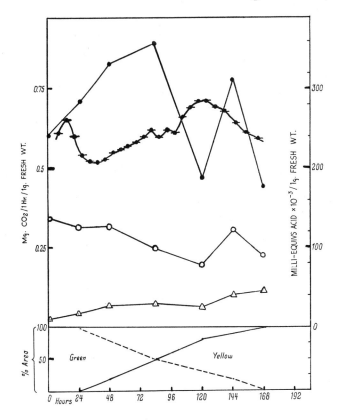

Fig. 20. *Changes of organic acids during starvation and the mean rate of carbon dioxide production. Yellowing of leaves shown for comparison. Carbon dioxide production,* ✦; *total titratable acid,* ●; *malic acid,* ○; *citric acid,* △ *(from Yemm and Somers, unpublished).*

equivalent, changes, and, if the cycle has been operative, the increase in "unknown" acids has been due to other components, or perhaps to small amounts of shorter chain fragments derived from carbohydrate breakdown or amino acid residues. There has been no retardation at step b, and, as we have seen before, the ammonia resulting from the amino acid oxidation has been fixed as glutamine.

The next period, from about the 12th hour onwards, was characterised by a fairly rapid fall in the respiratory quotient to about 0.8. Carbohydrate was no longer the major constituent consumed, and Yemm deduced that proteins were then being utilised in respiration. In the absence of a complete leaf analysis, such as that displayed in table 87, it is doubtful if such a deduction is warranted on chemical grounds, for large amounts of undetermined, yet consumable, products may have been present. The more recent analysis of the organic acids would, however, appear to confirm this deduction, for the malic acid loss was very little more than could be accounted for by an equimolar synthesis to citric acid, and the unknown acids must have been derived, at least in large part, from other sources (such as amino acid fragments). The initial "carbohydrate phase" had therefore been succeeded by a "protein phase," and the small accompanying organic acid metabolism had kept the respiratory quotient slightly above the theoretical value of 0.7. Moreover, the fact that the ammonia resulting from amino acid oxidation, in this second period, had gone over very largely to asparagine, suggests that the deaminised amino acid residues might have been, to an increasing extent, taken into the cycle at step d as succinic acid. In the final stages, ammonia was rapidly accumulating, and, to provide energy, oxalacetic acid or fumaric acid was being withdrawn again from asparagine. This was the beginning of an "organic acid phase" which would have taken the respiratory quotient above unity (compare fig. 6), but the concentration of ammonia soon became toxic, and death resulted.

I can mention briefly one other experiment with detached leaves in which the organic acids have been satisfactorily determined—that of Vickery, Pucher and their colleagues with rhubarb—and I have to thank Doctors Vickery and Pucher for permission to make use of their extensive (unpublished) data. The results are complicated by the fact that the leaves were stood with their long fleshy petioles in water or glucose solution throughout the experimental period, so that translocation of products from the blade to the petiole and vice versa was possible. When freshly picked, the general analysis of the leaf extract resembled that of tobacco leaves, i.e. the organic acid content was fairly large and the identified acids—malic, oxalic and citric—accounted in very large part for the ether-extracted organic acids. The two main points of interest are as follows. In the first place, the blades, irrespective of whether the petioles were stood in water or glucose in the dark, or in water in the light, showed, during a period of 165 hours, a progressive decrease on both malic acid and citric acid. Assuming that the cycle were operative, therefore, there was no impediment at step *b*, and the *only amide* elaborated at the expense of the ammonia derived from amino acid oxidation was glutamine. In the second place, protein breakdown, as in all the cases cited above, was initiated in the early stages before consumable carbohydrates had become depleted.

The interpretation that I have given to these five sets of starvation experiments is admittedly speculative, in that it is based on the hypothesis of an organic acid cycle in respiration for which there is no *ad hoc* proof at present, and perhaps there never can be. I have been encouraged in this policy by the wonderfully complete data of Vickery, Pucher, Wakeman and Leavenworth portrayed in table 87, which, for the first time, have given the chemist grounds for believing that no "unknown" groups of substances can be playing any major role in the metabolism of these tobacco leaves, and inspire faith in the carbon balance sheets which I have made use of in my discussions. And I would like to add that my reading of a very extensive, if scattered, literature during the preparation of this book has impressed me with a firm belief that it is the group of substances we refer to as "organic acids"—which have, in the past, been neglected in studies of plant metabolism because satisfactory methods of characterisation were not available—that occupy the central, and therefore the key position, in the carbohydrate, protein and fat metabolism of plant cells.

Much, however, remains to be learned, and it may even be argued that the slow accumulation of citric acid in the darkened tobacco leaves, and the failure to make use of this component of the cycle in the later stages of starvation, are good evidence against the cycle having been operative at all! The difficulty is admitted, and the tentative suggestion has been made that there may be some impediment, perhaps in the enzyme systems concerned with step b, which operates more and more with duration of starvation. In this connection, it is to be remembered that the tobacco leaves were making very extensive use of malic acid during approximately the first 70 or 80 hours in darkness, and it may well be that, in the later stages, the cycle was operating at a very slow rate, and asparagine was then being formed *via* succinic and fumaric acids (step d), without effective entry into the cycle. In three of the other cases cited, citric acid did not accumulate during the first 100 hours, and there was thus no tentative evidence against the operation of the cycle. As I have already pointed out, it is significant that, in each of these cases, glutamine was readily synthesised, suggesting that step b was being traversed without check. Further evidence is clearly called for, and methods of organic acid analysis must be developed so that the "unknown" fraction encountered, for instance, by Yemm and Somers, can be more definitely characterised. If, as I think, the slowest partial step in the tobacco leaf cycle was $b,$ leading to citric acid accumulation, then, in other species, it may well be that a different step will occupy this position, and it should then be possible to observe an accumulation of another acid characteristic of the cycle. It is on such circumstantial evidence that our judgment of this important question must be based, and one can only hope that agreement will be forthcoming in a sufficient number of cases to constitute, eventually, the strong presumption that a true verdict has been given.

Throughout the above discourse I have, in the appropriate places, emphasised the fact that, within a very few hours of leaf detachment, there is clear evidence of proteins having been metabolised: the significance of this observation is discussed in the chapter which follows.

CHAPTER XI

THE REGULATION OF PROTEIN METABOLISM IN LEAVES

MOTHES (178) appears to have been the first to consider seriously the factors that control the protein level in leaves. In the first place, certain experiments brought out clearly a relationship between proteolysis and water content. A deficiency of water in leaves attached to the plant was found to hasten protein decomposition and the translocation of soluble products, but the effect varied with the species and, in general, young leaves suffered less than old leaves. Further experiments, however, in which all but certain selected leaves had been removed from the plant, convinced him that the decisive factor was not the age of the leaf but its position on the stem and the consequent difference in suction power. If the leaves had been previously detached from the plant, proteolysis, under conditions of water deficiency, was slowed down in both old and young leaves, and he considered that this was due to the action of the accumulated soluble products on the enzymes concerned rather than to any disturbance of equilibrium. The actual rate of proteolysis, however, was faster in the young leaves than in the old ones, so that the increased proteolysis observed in old leaves when these still remained attached to the plant could be related only indirectly to leaf age. From these investigations, Mothes concluded that, in studies of protein level in leaves, the whole plant itself, as well as these organs, must be examined, for translocation of soluble products from the older parts to the new growing tissues was probably one of the important factors concerned.

In further amplification of these views, his pupil T. Schulze (302) investigated protein synthesis in detached leaves kept under various conditions both with and without nutrients. The results led him to the conclusion that—given adequate supplies of soluble nitrogen—leaves attached to the plant normally attain a protein level limited by a "stability value" which will depend on the development of the plant, and will be highest when the leaves are young. The differences in these "stability values," however, could not be correlated with translocation effects, nor with the competition between the upper and lower leaves for soluble products, and he ascribed them to the action of mutators (activators and paralysers) on the proteolytic enzymes concerned. In support of these ideas, he prepared glycerol extracts of leaves detached from the plant at different stages of development and showed that, in varying degree, they possessed towards casein and edestin proteolytic activity which was amenable to treatment with extraneous "activators" such as glutathione, cysteine, hydrogen cyanide, etc. Acetone extracts of these leaves, on the contrary, were free from enzymes, but they contained what he called "mutators" which, under certain conditions, could activate or paralyse the enzymes present in the glycerol extracts. The "mutators" were "redox" substances of unknown composition, and their effect, which was ascribed to an intensification of oxidative processes, was strongest when they were prepared from leaves which, before detachment from the plant, were exhibiting pronounced protein decomposition, e.g. during the flowering period.

In a later review of this work, Mothes (180) stated that oxygen was probably the decisive factor in leaf protein metabolism, in that the oxygen potential regulated (subject to pH control, etc.) the activity of the proteolytic enzymes concerned. Thus a high oxygen potential favoured protein synthesis, as was shown by the fact that, in leaves attached to the plant and kept in an atmosphere of pure oxygen, the normal decrease in protein content at night was not observed (he was assuming here a diurnal rhythm in the protein content of leaves—see below) while a low oxygen potential, as in detached leaves deprived of oxygen, led to very rapid protein decomposition and death of the cells. Furthermore, the suggestion was in keeping with the known correlation between respiratory activity and protein synthesis, for both required a high oxygen potential or tension.

These views have been caustically criticised by Paech (200). In a well-reasoned statement, supported by the necessary *ad hoc* experiments, he showed, in the first place, that under conditions of anaerobiosis leaf cells were rapidly killed, and that before death, little or no protein breakdown had occurred, so that the intense proteolysis noted by Mothes must have been due to post-mortem changes. Secondly, he pointed out that, as soon as the living structure of a plant cell was destroyed, the proteases were readily attacked by oxygen, and their activities were so changed that, in autolytic experiments (as *per* T. Schulze), the intensity of the proteolytic action was reduced by the oxygen tension of the surroundings. His own experiments showed that this influence of oxygen on proteases did not occur *in vivo* and it was clear, therefore, that the suggestions of Mothes were not applicable to the explanation of protein regulation in the living leaf cell.

Paech's views on this problem were based, in large part, on an extensive review of the literature, which led him to the tentative conclusion that the protein level in a plant cell must be determined by its content of both total nitrogen and monosaccharides, variations in these factors leading automatically, by mass action, to synthesis or decomposition.

He dealt first of all with the influence of carbohydrate supply, and cited, *inter alia,* the early work of Suzuki (319, 321) and Zaleski (368) showing that protein synthesis in leaves could be brought about by feeding with sugars. Next, he drew attention to the work of Deleano (see pp. 211 *et seq.*), who showed that, when detached vine leaves were kept in the dark for about 100 hours, the only noticeable change was a loss of starch; no protein was decomposed and the concentration of monosaccharides present remained very like that in illuminated leaves. On prolonging the period in darkness, however, protein loss occurred, and this was accompanied by a parallel fall in the monosaccharide concentration. He admitted, nevertheless, that other workers, e.g. Mothes (176), had been able to detect protein decomposition in leaves within quite a short time—10 to 20 hours—after detachment from the plant, and to throw further light on the part played by carbohydrates in this connection, he carried out some experiments of his own.

Intact, but not too old, leaves were first kept in the dark until they were, as yet, in no danger of yellowing, but showed a noticeable amount of protein decomposition.

TABLE 89.
(From Paech, 200.)

Leaves of *Phaseolus multiflorus*. Temp. 24°	Protein-N in percentages of total leaf-N	Leaves of *Helianthus annuus*. Temp. 24°	Protein-N in percentages of total leaf-N
Analysed at once	85.5	Analysed at once	93.3
After 95 hours in dark	80.8	After 96 hours in dark	69.3
Ditto, then twice infiltered with 2% glucose and kept another 24 hours in dark	83.6	Ditto, then infiltered twice with 3.5% glucose and kept another 24 hours in dark	70.7
Leaves of *Brassica napus*. Temp. 24°		Leaves of *Lupinus luteus*. Temp. 24°	
Analysed at once	84.8	Analysed at once	90.0
After 85 hours in dark	69.0	After 66 hours in dark	77.7
Ditto, then infiltered twice with 4.5% glucose and kept another 9 hours in dark	72.6	Ditto, then infiltered 3 times with 3% glucose and kept another 35 hours in dark	77.2
		Ditto, then kept another 35 hours in dark without infiltration	71.0

Then, using the half leaf method, one half was analysed immediately and the other half after a further period in darkness following infiltration with glucose solution.

These results showed that an increase in the supply of carbohydrate could either arrest protein breakdown, or even bring about protein synthesis, and were thus in keeping with the earlier work of Mothes (176) in which detached leaves had been cultured on glucose solution. Protein breakdown, therefore, depended on the carbohydrate level, and he cited Borodin (see p. 31), Hansteen (106) and Björkstén (16) as authority for the dictum that only monoses could produce this effect.

TABLE 90.

(From Mothes, 176.)

Detached and darkened prophylls of *Phaseolus multiflorus* kept with petioles in	Protein-N in percentages of total leaf-N		Change in protein content %
	At beginning of experiment	At end of experiment	
Water, for 4 days	81.7	64.7	—20.8
1% glucose, for 6 days	82.5	70.5	—14.5
2% glucose, for 6 days	86.7	79.3	—8.5
2.5% glucose, for 5 days	77.2	73.2	—5.2
4% glucose, for 4 days	80.9	80.5	—0.5
5.7% glucose, for 5 days	81.6	83.4	+2.5

Nitrogen was considered to be available for protein synthesis if it were present in the form of ammonia, amides, amino acids, protein bases and certain ill-defined "rest-compounds," and, again relying on evidence culled from the literature, Paech declared that protein synthesis in leaves, given adequate carbohydrate supplies, was proportional to the "active" nitrogen (potential ammonia) available. He cited *inter alia* the work of Hansteen (106), who fed asparagine and urea, of Saposchnikow (265), who fed ammonium salts, and the infiltration experiment of Mothes quoted in table 69, page 198, showing that well-nourished leaves of low protein content could synthesise protein from ammonium malate.

From considerations such as these Paech felt he was justified in giving a more precise expression to the above-mentioned mass-action effect, and suggested that, in the intact plant cell, the control towards protein synthesis or decomposition would depend on the amounts of chemically active forms of carbohydrate (monoses) and of chemically active forms of nitrogen (ammonium salts, amides and to a lesser extent amino acids and bases), in that an increase or decrease of the component present *in lesser amount* would bring about equivalent protein changes. The upper limit of what he called this "mass action law" was to be found in the maximum storage power of the cell for protein, and the lower limit in the minimum amount of protein necessary for the maintenance of protoplasmic structure; these would not be narrow limits, and the amount of protein required for the lower limit might be surprisingly small. Such a hypothesis required that large amounts of active monoses and active nitrogen should never be present together in the same cell or in the same sphere of action, and as soon as a suitably high concentration of both components was available—*regardless of any other conditions, such as stage of development, protein concentration, light or dark, etc.*—protein would be synthesised.

Having shown that a lowering of the sugar level reduced the protein content of the cell, that feeding monoses caused protein to be formed if suitable soluble nitrogenous products were present, and that in the presence of ample monose the mere supply of ammonium salts was sufficient to cause protein synthesis, it remained to be proved that withdrawal of nitrogen compounds would lead to protein decomposition. The experimental difficulties here were not easily overcome, for one cannot remove soluble products from excised leaves, and he finally made use of seedlings, which show a unidirectional translocation from the reserve organs to the meristematic tissues.

The experiments have been reported fully in Chapter III, pages 67 *et seq.*, and I need only add here that the results were in accordance with Paech's expectations.

Paech next considered the withdrawal of nitrogen from the older leaves of a plant during the intensive growth of the flowering period, when nitrogen requirements would normally no longer be covered by root intake from the soil. He assumed here, as in his seedling experiments, that the growing parts must exert some "attractive force"—which he could not explain—on the soluble nitrogenous compounds of the older leaves, leading automatically to protein decomposition therein, which would preserve the constancy of the relative protein value (protein nitrogen: total nitrogen) observed by Smirnow (311) and Gowentak (95) in their studies of the seasonal variation in the nitrogen content of leaves. Finally, he crystallised his views on protein metabolism in the plant with the statement that "in any single organ—within the limits of its cell capacity—the protein content is the resultant of the quantity of monosaccharides and soluble nitrogenous materials ($= NH_3$) present, and that on the part of the whole organism these single systems are linked together, being balanced against each other by means of the translocation stream whose direction and velocity is determined by the forces exerted at the growing centres." In applying this mass action principle he assumed, of course, that the other elements required for protein synthesis, e.g. phosphorus, sulphur, potassium and magnesium were also present in optimum concentrations; if not, their deficiency would form a limiting factor.

Paech has collected a vast array of data—mainly culled from the work of others—in support of his views, and in so far as his interpretation is confined to these particular data one can, with certain reservations concerning chemical mechanisms, agree with many of his conclusions. But Paech himself has gone further than this, and has propounded a "mass action law" which he considers applicable to all plant cells, so that his views demand, and must receive, a more searching enquiry.

My immediate criticism of his paper is levied against the curious and facile way in which the protein-sugar balance in seedlings and leaves has been assessed: on the one hand, we are presented with an accurate estimation of the amount of protein present, variations of 2–3 per cent being regarded as significant and taken as valid evidence in support of the "law"; on the other hand—in strange contrast—no corresponding sugar analyses whatsoever are quoted, either from his own or from the work of others, and the sugar levels have been judged by external conditions (light, dark, supply of extraneous sugar, etc.). The literature, at the time, provided very little *ad hoc* data for use in this connection, and for this reason alone Paech himself should have investigated the variations in the sugar content of his experimental material.

TABLE 91.
(Unpublished data of Vickery, Pucher, Wakeman and Leavenworth.)

Tobacco leaves cultured on water in light
Series 1935 — Per 1000 g. fresh leaves

Hours	0	24	74	120
Protein-N	1.69	1.71	1.32	1.11
Sucrose	0.29	1.07	2.09	1.83
Glucose (fermentable sugar − sucrose)	0.76	2.10	6.15	7.19
Unfermentable sugar	0.43	1.13	2.97	3.46

Yemm's results (fig. 4, p. 213, and fig. 7, p. 217) show at once that the "law" is not applicable to barley leaves. The total nitrogen was unchanged, yet during the first few hours there was a fall in protein content accompanied by a rapid increase in the amount of glucose present. Equally convincing evidence is provided by some (unpublished) experiments of Vickery, Pucher, Wakeman and Leavenworth, in which tobacco leaves were cultured on water in continuous light (table 91).

In this case the protein breakdown was accompanied by a steep rise in the glucose content, to a value of more than 5 per cent of the total dry weight.

Although Paech has correlated the protein and soluble nitrogen with monoses, he states specifically that the immediate nitrogen-free precursors of the amino acids required for protein synthesis are probably a series of α-ketonic acids, and that it is these acids which must first be formed, in some unknown way, from the monoses. It seems to me that the weakness of his suggested relationship lies in the assumption that it is the active mass of the monose present in the plant cell which determines the active mass, each and severally, of these α-ketonic acids. In some instances, as his own experiments show, the assumption is not at variance with observation, but in others, as we have just seen, the contrary conclusion holds. If mass action does, indeed, function as a control in protein synthesis and degradation, then I would tentatively suggest that it is the complete series of α-ketonic acids, and not the monoses, which stand in the postulated relationship with the soluble nitrogen and protein. It is to the credit of Paech that he realised, far more clearly than those who had previously investigated the question, that protein synthesis must depend on the level of some source of carbon as well as of nitrogen, and his contention that the reaction is controlled by the component present in lesser amount is certainly in keeping with many observations, as we saw in Chapter III. In spite, therefore, of my firm conviction that Paech was mistaken in attributing any predominant role in protein regulation to the active mass of monose present, I think his views may still be of great use to us if his suggested relationship be amended to read as follows: "In the intact plant cell, the control towards protein synthesis or decomposition will take place through the amounts of a necessary series of α-ketonic acids and of active nitrogen (ammonium salts, etc.), in that an increase or decrease of the component present in minimum amount will bring about equivalent protein changes." Since we do not yet know in what manner and under what conditions these ketonic acids are synthesised in the plant, the amended relationship, unlike the original, has the additional merit of emphasising the complexity of the problem and not its simplicity.

It will be realised that all these suggestions of Paech imply that the proteins of leaves are stable substances, and that protein decomposition or synthesis will take place only when variations occur in those simpler products with which they stand in equilibrium controlled by mass action law. Contrary views, however, are held by other workers.

We saw in Chapter II that, as early as 1878, Schulze (see pp. 40 *et seq.*) had mooted the notion that asparagine accumulation in the meristematic tissues of seedlings might be due to a continuous breakdown and rebuilding of protein, and that in the following year Borodin (see p. 27) had stated that some such process might be responsible for the respiration of plant cells, carbon dioxide being produced during protein regeneration at the expense of asparagine and monoses. Such ideas, however, were soon forgotten or abandoned, and it was only in recent years that suggestions reminiscent of them have again been tentatively put forward.

In 1923, from a review of the older literature (32) and also from my own experiments with bean leaves (basis of comparison: fresh weight and twin leaf) (35), I deduced that there was a fall in the protein content of leaves at night, which I ascribed (37) to a diurnal variation in the respective rates at which a continuous protein synthesis and decomposition took place in these organs. This purely speculative assumption was based on evidence the validity of which has not been admitted by all subsequent investigators, for it is difficult to provide a really satisfactory basis on which such a small diurnal change in protein content as that observed (*circa* 2 per cent) can be measured with any degree of certainty. The first experiments of Mothes (176), who used many different species of leaves (basis: fresh weight) and the experiments of Mason and Maskell (155) with the cotton plant (basis: residual dry weight), appeared to confirm a diurnal fall in protein content at night, but Gouwentak (95) (basis: unit of leaf area) and Denny (64) (bases: half leaf, twin leaf and residual dry weight) could find no significant change.

Mothes (178), in a more recent investigation, has shown that the effect depends on the stage of development of the plant. Using *Nicotiana sp.* (basis: half leaf) he found that young leaves increased their nitrogen content (protein nitrogen was not separately determined) both by day and by night, mature leaves exhibited a slight loss by night and gain by day, while old leaves showed a decrease both by night and by day. All the leaves showed an increase in fresh weight during the night, and younger leaves an increase in area, due to growth, emphasising the need of measuring changes in absolute units, as in the twin and half leaf methods. In leaves of the shrub *Syringa vulgaris* there was, during the vegetation period, a slight decrease in both total and protein nitrogen during the night, and correspondingly unimportant increases during the day, but, in the autumn, there was an intense drainage of nitrogen from the leaves, and the protein level fell both during the day and at night. On the whole his results were in agreement with those of Mason and Maskell, and he concluded that it was only during the active vegetative period that there was a diurnal rhythm in the protein content of leaves, when the effect was ascribable to the translocation stream. It would appear from discussions, and a later review (180) that Mothes considers that, in leaves, protein synthesis and decomposition are controlled by separate enzyme systems, and that he regards both processes as continually at work.

More concrete expression has recently been given to the view that a protein cycle is operative in leaves by Gregory, Richards and their collaborators who have approached the problem of protein synthesis from an entirely different angle from that of the investigations discussed previously. Their researches, which have continued without intermission since 1921, always with a pure strain of barley, began with a study of the factors controlling growth, primarily with a view to analysis of the effects of various nutrients. The attempt to characterise the specific effects of various nutrient elements led to a study of respiration, and, to elucidate the results obtained along this line, they were led to investigate the carbohydrate and nitrogen metabolism of their plants. In striking contrast to the type of data with which many of the previous investigators were satisfied, Gregory's work has attempted to gain a wider basis for generalisation by systematically studying the changes in each successive leaf throughout the growth cycle. Thus the conditions of nutrient supply and the type of plant material studied were standardised, and results therefore could be compared year by year, and in this way a unique body of data has accumulated. Again, the statistical methods of Fisher have been applied throughout in the analysis of the experimental results, so that the reliance to be placed on the data is given objective expression.

It would lead far beyond the scope of this work to consider in any detail the relations established between the processes of respiration, carbohydrate and nitrogen metabolism in successive leaves of plants supplied with varying levels of the nutritive elements; nor is this necessary since Gregory has summarised the information obtained in a recent review (100). Suffice it here to say that the metabolism of the normal plant has been compared at each corresponding stage with that of plants deficient to varying degrees in the essential elements of nitrogen, phosphorus and potassium. Only in so far as the abnormal metabolism of such plants throws light on the related processes of respiration, and of carbohydrate and nitrogen metabolism, will they be here of any immediate concern.

The type of data, then, to be considered is as follows.

Respiration measurements, and data of carbohydrate and nitrogen analyses, obtained at successive definite stages during development, are available, showing simultaneously the changes in respiration rate and the drifts of carbohydrate and nitrogen metabolism. The successive leaves constitute a series each member of which differs in the levels of carbohydrate and various nitrogen fractions present; and thus a wide range is obtained without the artificial resort to infiltration or feeding. The nutrient deficiency series are useful, since by their aid the relative levels of carbohydrate and nitrogen fractions may be greatly changed.

In a preliminary study, Richards and Templeman (240) examined the successive leaves of the main shoot at full emergence and during senescence, from plants deficient in nitrogen, phosphorus and potassium. In general the total nitrogen and most of the nitrogen fractions reached a maximum in the second to fourth leaf, declined to a minimum in the eighth or ninth and again rose in the last leaf. Their results with respect to nitrogen and potassium deficiency have been more recently confirmed and amplified by Gregory and Sen (102), and to phosphorus deficiency by Richards (239).

In the fully manured plants, a close parallelism was found between the nitrogen metabolism and respiration in the successive leaves at emergence, which was, moreover, independent of any carbohydrate supply. The same general relationship also held in the leaves of nitrogen deficient plants; both respiration and the nitrogen level were low, but the relative concentrations of the various nitrogen fractions remained unchanged during the leaf succession, indicating a normal course of protein metabolism.

In the potassium-starved leaf (with high sodium, low calcium), however, conditions were very different. Here the protein was, at the time of emergence of the leaves, at least as high as in the normal leaf, the mean protein level in all leaves being the same as in fully manured plants.

TABLE 92.

(From Gregory and Sen, 102.)

Mean values of respiration rate and analytical data for the various manurial series.

(Respiration rate in mg. per g. dry weight per hour; other data as percentage of dry weight.)

Treatment		Leaves included	Respiration	Free reducing sugar	Sucrose	Protein-N	Amino-N
Control N_1K_1		3–10	6.00	2.02	14.13	3.27	0.273
K-deficient	N_1K_3	3–10	7.86	1.58	11.11	3.32	0.379
	N_1K_5	3–10	7.46	1.37	4.10	3.32	0.464
N-deficient	N_3K_1	3–9	4.49	1.21	22.14	1.70	0.148
	N_5K_1 and						
	N_5K_5	3–7	4.13	1.38	19.02	1.25	0.098

$N_1, N_3, N_5 =$ levels of nitrogen $(1, \frac{1}{9}, \frac{1}{81})$
$K_1, K_3, K_5 =$ levels of potassium $(1, \frac{1}{9}, \frac{1}{81})$

TABLE 93.

(From Gregory and Sen, 102.)

Protein-N content
(percentage of dry weight)

Leaf No.	Control	Increasing K deficiency		Increasing N deficiency		N- and K-deficient
	N_1K_1	N_1K_3	N_1K_5	N_3K_1	N_5K_1	N_5K_5
3	3.23	3.50	3.58	3.13	1.81	1.58
4	3.87	3.78	3.79	2.39	1.34	1.35
5	3.94	3.81	3.66	1.84	1.39	1.42
6	4.06	3.42	3.52	1.29	0.93	0.83
7	3.70	3.42	3.35	1.02	0.90	0.92
8	2.58	3.08	3.18	1.06	1.06	1.19
9	2.29	2.73	2.76	1.18	–	–
10	2.52	2.84	2.75	1.69	–	–
Mean	3.27	3.32	3.32	1.70	1.24	1.21

$N_1, N_3, N_5 =$ levels of nitrogen $(1, \frac{1}{9}, \frac{1}{81})$
$K_1, K_3, K_5 =$ levels of potassium $(1, \frac{1}{9}, \frac{1}{81})$

At the same time there was a marked increase in the amount of amino and amide nitrogen (240), both of which appeared in normal proportions, accompanied always by a low sugar content and an increased respiration. In the later leaves, moreover, nitrate accumulated.

TABLE 94.

(From Gregory and Sen, 102.)

Total amino-N content
(percentage of dry weight)

Leaf No.	Control	Increasing K deficiency		Increasing N deficiency		N- and K- deficient
	N_1K_1	N_1K_3	N_1K_5	N_3K_1	N_5K_1	N_5K_5
3	0.341	0.338	0.466	0.249	0.126	0.110
4	0.383	0.417	0.464	0.238	0.113	0.108
5	0.371	0.418	0.437	0.169	0.112	0.095
6	0.288	0.409	0.482	0.108	0.073	0.069
7	0.278	0.326	0.464	0.099	0.096	0.079
8	0.218	0.444	0.415	0.080	–	0.086
9	0.165	0.380	0.559	0.096	–	–
10	0.138	0.302	0.427	–	–	–
Mean	0.273	0.379	0.464	0.148	0.104	0.091

N_1, N_3, N_5 = levels of nitrogen (1, ⅑, ⅟₈₁)
K_1, K_3, K_5 = levels of potassium (1, ⅑, ⅟₈₁)

The development of these plants with low potassium supply (⅑ normal) suggested that protein synthesis went on very rapidly: a supernormal tiller production was found, leaf production rate was normal and the leaves were of normal size, indicating that meristematic activity in such plants was very high. Richards and Templeman (240) therefore suggested that potassium was not primarily associated with protein synthesis, but that it was necessary for maintaining the protoplasmic complex and in its absence rapid proteolysis occurred. In potassium-starved plants protein was, in fact, very unstable; very rapid death of the individual leaves occurred and there was a rapid fall in protein level during senescence.

Gregory and Sen (102) realised the necessity of bringing all these experimental results into some relationship with current views on the respiratory process, and in an endeavour, however tentative, to discuss the interrelation of carbohydrate and nitrogen metabolism in determining the levels of respiration found in the successive leaves within a series, as well as the differences in respiration level between the various series, they put forward the following schema as a basis of discussion.

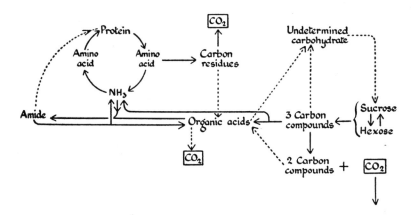

They admitted that, in attempting to correlate the respiration of carbohydrate and protein synthesis in this way, they had leaned rather heavily on views expressed by specialists in these fields. Moreover, it must be remembered that their idea of linking the metabolic cycles of carbohydrate and nitrogen through organic acids in this interesting way was formulated before any suggestions had been put forward that an organic acid cycle itself might function in respiration.

In support of this schema, they discussed first of all two alternative views on the relation between protein synthesis and degradation. If protein synthesis were the reverse of proteolysis, then the direction of the reaction would be determined by mass action and the velocities of the reactions concerned. The accumulation of amino nitrogen in the potassium-starved leaves, therefore, should have tended to the reverse action and the levels of amino nitrogen and protein should have varied in proportion. This might have been countered by rapid translocation away of amino nitrogen, but this, on the other hand, would have led to a relatively low ratio of concentration of amino nitrogen to protein. In fact the reverse appeared, and an accumulation of amino nitrogen was seen. This suggested that the route of protein synthesis might be different from that of protein degradation, and that the rates of these processes might vary independently, and in fact that a cycle was operating. Variation in velocity of the reactions concerned would lead to accumulation: thus if the synthetic route to protein were rapid, protein would tend to accumulate, and by mass action increase the breakdown to amino acid. If, furthermore, as in the schema, these amino acids, before being again used in protein synthesis, were deaminated, and if this particular reaction were the slowest in the cycle, then amino nitrogen would accumulate.

Gregory and Sen (102) suggested that the very high respiration rate of the potassium-starved leaves was undoubtedly associated with the accumulation of amino nitrogen, for such a view was in keeping with the earlier work of Spoehr and McGee (314) and Schwabe (305). In comparing manurial series at high, medium and low potassium (K_1 K_3 K_5), it was found that, for comparable successive leaves after the third, the respiration rate of the medium series was greater than either extreme, as Richards (238) had first shown, and they had confirmed. They pointed out that the theory of preferential oxidation of carbohydrate was closely concerned in this connection. Gregory and Baptiste (101) had found a low level of both reducing and total sugars in these leaves and considered that it was associated with the low assimilation rate, high respiration, active protein synthesis and excessive meristematic activity. Feeding with potassium alone, however, had no effect on the respiration of such leaves (detached from the plant) (Said, 263), but, if given with sugar, an immediate increase in respiration resulted showing that oxidative deamination could not be alone concerned. If, Gregory and Sen remarked, carbohydrate was preferentially respired, while sugar supply in potassium deficiency was low through failure of photosynthesis, how, on this assumption, was protein synthesis maintained at a level at least as high as in the fully manured leaf, in spite of low carbohydrate concentration?

Clearly, other factors were involved besides the level of carbohydrate, and they suggested that, in the potassium-deficient leaf, conditions were somewhat as follows. Rapid synthesis of protein occurred, and at the same time rapid protein degradation, the two sets of processes being distinct. The amino acids resulting from the protein degradation underwent deamination to give ammonia, which then condensed with products—possibly organic acids—derived from carbohydrate to give amides. The carbon residues of deamination might perhaps have been completely respired away. From amide, by a different route through amino acid, or directly from amide, protein was resynthesised. In the potassium-deficient leaf, protein level was normal and amino nitrogen much increased; the protein cycle must therefore have been traversed more rapidly than normally. The output of carbon dioxide was thus related to the more rapid rate at which the cycle was traversed. This same hypothesis would equally well account for the situation in the nitrogen-starved leaf. Here the cycle was very slow, and the output of carbon dioxide low in spite of high concentration of sugar. In the fully manured plants, during the progress through the successive leaves, protein metabolism declined, and, in step with this, the respiration fell in spite of accumulation of sugar in the later leaves. Tentatively, therefore, Gregory and Sen put forward the suggestion that carbohydrate oxidation in the green leaf was regulated by the rate of protein synthesis. The production of carbon dioxide might or might not be a resultant, in the living leaf, of the process of deamination, but even if carbohydrate respiration was the main source of carbon dioxide output, the intermediate products being utilised through organic acid formation in protein synthesis, yet there was a general "regulation" at work which controlled the whole process, though the nature of this was still obscure.

In amplification of this work, Richards (239) has recently shown that, under conditions of phosphorus deficiency, a remarkably close and simple relationship held throughout the whole series of successive leaves of plants, maintained at various phosphorus and potassium levels, between respiration rate and protein content after slight

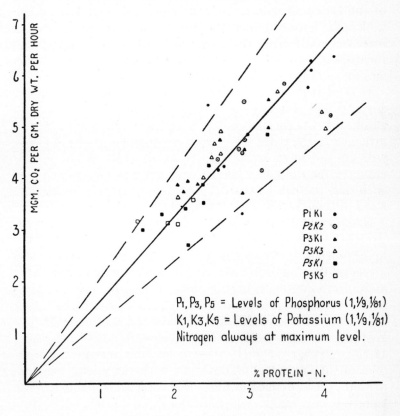

Fig. 21. *From Richards, 239.*

adjustments had been made for the level of other factors. The individual values for the successive leaves in all the series are shown in Fig. 21: the calculated regression line is given, together with the limits of significant departures from the expected values.

To explain this reciprocal relationship, it was suggested that a given rate of carbon dioxide evolution could maintain only a definite quantity of protein. Regarded from the standpoint of efficiency, this would be highest in the fully manured leaves, for here the loss of carbon as carbon dioxide per unit of protein present was least. Under conditions of phosphorus starvation, the importance of the reducing sugar concentration had been clearly established; here apparently the level of phosphorus in the leaf was controlling the mechanism of carbohydrate respiration in supplying the nitrogen-free precursors of the amino acids for further protein synthesis.

It is of very great interest to observe that the general conclusions which Gregory, Richards and their collaborators draw from their comprehensive studies with leaves attached to the plant show certain remarkable similarities with those I have deduced from an analysis of the chemical data for detached leaves presented in Chapter X; in each case the role of the proteins in respiration is emphasised, although not to the same degree.

Gregory and Sen postulate that a protein cycle is continuously at work in leaves and that the paths of protein breakdown and of protein synthesis must be separate, the nitrogen-free amino acid residues being respired, and amino acid synthesis taking place from new carbon chains derived from sugar. Examined in light of reactions 6 and 13, we see at once that such a thesis can be considered in keeping with the chemical mechanisms portrayed therein. If the purpose of such a cycle be to regulate the production of carbon dioxide from both carbohydrate and protein, then the amino acids resulting from protein degradation must undergo oxidative deamination according to reaction 13, and the nitrogen-free residues must either enter the cycle at stages d or e, being subsequently metabolised to carbon dioxide, or must provide for the synthesis of amides, with or without effective entry into the cycle, at stages f, g and h. In either case, these nitrogen-free products will lose all previous generic relationships, so that a resynthesis of protein according to reaction 6 must take place either at the expense of the amides, or from a new series of a-ketonic acids metabolised from sugar. We can even go one step further with Gregory and Sen and suggest that the extent to which the protein itself enters directly into respiration will be a function of amide synthesis: the greater the ease with which this occurs the more will carbon dioxide be produced from sugar.

All such ideas, however, proceed on the assumption that a protein cycle is operative in leaves, and the evidence for this is, at present, purely circumstantial. Gregory and Sen make effective use of their data for potassium-deficient leaves, and although the chemist could suggest alternative explanations for the accumulation of amino nitrogen, these would not necessarily satisfy the numerous other observations they bring forward in support of their thesis. Then, again, we have the many instances cited in Chapter X which show that, in detached leaves, it is often possible to observe protein breakdown, with accompanying oxidative deamination of the resulting amino acids, within a few hours of detaching leaves from the plant. Can we regard this as evidence in favour of a protein cycle, the mechanism of protein synthesis having become impaired, so that protein degradation predominates? It is possible, but in my opinion conditions here are too complex for a definite answer at the present time. We know, for instance, from the work of Deleano, of Mothes and of Yemm, that the rate at which protein breakdown occurs in detached leaves kept in the dark varies from species to species, from leaf to leaf with age for the same species, and also with the carbohydrate level. Such observations could, indeed, be used as evidence that the rate at which the protein cycle is operative in plants varies from species to species or from leaf to leaf, for we know, from the data of Vickery and his colleagues given in table 91, that the carbohydrate level cannot be the decisive factor in controlling protein breakdown.

Gregory and Sen (102) have, in fact, calculated the maximum rate at 25° at which protein breakdown might take place in barley leaves attached to the plant. On the hypothesis that all the carbon lost as carbon dioxide might have had its origin in protein, they calculated the amount of protein breakdown necessary to account for the carbon dioxide output. Using the data in table 92, and selecting the series with highest mean respiration rate (N_1K_3), the mean protein nitrogen content was 3.32 per cent of the dry weight, equivalent to approximately 20 per cent protein, or 100 mg. carbon capable of producing 370 mg. carbon dioxide per g. dry weight. As the mean respiration rate was 7.86 mg. per g. dry weight per hour, this corresponds to a loss of 2.1 per cent of the protein content per hour. Even without taking into account protein synthesis, which might have been occurring concurrently, Gregory and Sen considered that this value was not incommensurate with those actually found during proteolysis in the darkened leaf. This is true, for the data of Vickery and his colleagues for tobacco leaves cultured on water both in light and in darkness showed a 25 per cent protein loss in 73 hours, or about 0.3 per cent per hour. If it be assumed that a protein cycle were operating in these leaves at the same speed as in barley, and that the normal rate of protein degradation had been maintained after detachment from the plant, then this loss would suggest that the rate of protein synthesis had by some means been decreased by one-seventh.

All of the points mentioned in the above discussion could be enlarged upon, but at the present time I think without profit, for the overriding fact remains that we do not yet understand the reason why, when certain leaves are detached from the plant, protein decomposition can be detected within a few hours. If this is indeed due to interference with the mechanism of protein synthesis then one is tempted to suggest that some influence of the root system, possibly hormonic, is responsible for the regulation of the protein level in leaves. On this important question, however, I would prefer to reserve judgement until I have had the opportunity to investigate detached leaves which have been induced to form adventitious root systems.

APPENDIX I

The Impurities Present in Leaf-Protein Preparations and Their Bearing upon the Estimation of Some Amino Acids, Particularly in Relation to the Formation of Humin during Acid Hydrolysis.

By J. W. H. Lugg

AS every student is well aware, the best that we can hope to achieve in chemical analysis is a mean of many estimations and a measure of its probable exactness. But our reliance on any analytical result obtained, whether it be an individual value or the mean of many, is necessarily limited at the outset to the degree of certainty that all systematic errors have been eliminated either by mutual cancellation or by individual correction to zero, and it becomes a matter of the most fundamental importance to examine the possible sources of these systematic errors.

Perhaps no field of analytical work is more subject to these errors than the biochemical. The materials submitted to analysis so frequently contain substances the composition and reactions of which are utterly unknown and which may well interfere with the estimation; the particular substance to be estimated is frequently of so labile a character that it cannot readily be separated from some of the possible interfering substances before the end reaction employed in the estimation (e.g. colour, precipitation, titration) is applied, and, finally, the limitations of the end reaction itself may be so ill-defined in consequence of the empiricism with which it is frequently loaded, that even when the substance to be estimated is initially in the free state the errors may be enormous. And if the substance is not initially free but must first be liberated from existing combinations, there is further scope for error. All this should be common knowledge, and indeed it may be, but a clear recognition of its implications is far from general, to which fact the biochemical literature bears eloquent testimony.

In estimating the amount of an amino acid residue in a protein preparation, the usual procedure is to hydrolyse the preparation by heating it in strongly acid or alkaline solution and to apply to the resulting hydrolysate a method believed to be satisfactory for the estimation of the particular amino acid in the presence of the other products of hydrolysis. The conceivable sources of error are many and attention has been drawn to them from time to time. It has been shown (Lugg, 144, 145) how the errors may be grouped for simplicity and as an aid to investigation into four categories: (a) those due to incomplete liberation from existing combinations, (b) those due to modification or destruction during liberation, (c) those due to modification or destruction subsequent to liberation, and (d) those properly associated with the imperfections of the end reaction employed. Each category may be investigated at least partially, but investigation of (a) and (b) is rendered particularly difficult by the existing dearth of complex synthetic peptides and complete absence of synthetic proteins of known composition; handicaps which are not in the least escapable by recourse to methods of estimation in which the plain hydrolysis step can be omitted. For the fact remains that an amino acid residue in a protein is not the same thing as a molecule of the free amino acid.

These points will be illustrated by describing the difficulties which have attended the estimations of some amino acids in the impure leaf proteins (Lugg, 146a and b).

Estimations of Cystine and Methionine.

Pollard and Chibnall (220) hydrolysed a selection of leaf-protein preparations with hydrochloric acid, freed the hydrolysates from insoluble acid-humin (fairly large amounts of which were produced) and estimated the cystine in the hydrolysates by Prunty's method (232), which is based upon the highly specific Sullivan (318) reaction. The results varied greatly with the different preparations. They believed the main organic impurity in the preparations to be a polysaccharide in an amount generally representing some 15 to 25 per cent of the weight of a preparation, and, knowing that quite small amounts of cystine (in contradistinction to cysteine) are lost when heated in acid solution with such quantities of sugars and cellulose (Lugg, 144), they considered their values to be fairly reliable on the supposition that the reduced form (cysteine) did not occur in the proteins.

It was found later that large fractions of the total sulphur appeared in the insoluble humin. The probable origin of this humin-sulphur was revealed in some experiments made by Bailey (7). He had found that all but several per cent of the sulphur in crystallised edestin could be accounted for as cystine and methionine by hydrolysing the protein with hydrochloric acid and applying to the hydrolysate reasonably specific methods for the estimation of cystine (Lugg, 142, 143) and methionine (Baernstein, 5). Minute amounts of sulphur were lost during hydrolysis. On adding 10 per cent by weight of arabinose to the protein before hydrolysis, Bailey encountered a loss of 11 per cent of the sulphur in the humin precipitated in the region of its minimum solubility (about pH 2.5), and overall losses of 30 per cent of the cystine and 20 per cent of the methionine. Thus, whilst nearly 90 per cent of the sulphur remained in the solution, much of it no longer belonged to unmodified cystine and methionine. Of the sulphur present in an impure leaf-protein preparation, as much as 30 per cent may be found in the insoluble humin after acid hydrolysis, and the overall losses of cystine and methionine are therefore likely to be very much greater than 30 per cent. At all events, the values obtained must be open to the gravest suspicion.

There are other methods of obtaining information as to the distribution of sulphur in a protein, however, and when these are employed with "pure" proteins such as crystallised edestin, they give results in fair agreement with those obtained by the methods already discussed. Baernstein (6) in 1936 published a procedure based upon the following reactions: The solid protein is hydrolysed in concentrated hydriodic acid solution at the boiling point, sulphate sulphur being reduced to hydrogen sulphide and sulphur dioxide, the cystine being reduced to cysteine, and the methionine yielding methyl iodide and the thio-lactone of homocysteine (by loss of water from homocysteine). The sulphur dioxide and hydrogen sulphide may be collected, oxidised and estimated as sulphate, while estimation of the methyl iodide evolved gives a measure of the methionine. The cysteine may be titrated in acid solution with iodate in presence of excess hydriodic acid, and the thio-lactone ring may be opened in mildly alkaline solution and allowed to reduce tetra-thionate to thio-sulphate, which, after re-acidifying the solution may in turn be titrated with iodate.* Blumenthal and Clarke (19), in their studies of the sul-

---

* The various reactions have been studied recently by Kassell and Brand (115).

phur distribution of various protein preparations, made use of the fact that cysteine may be made to yield its sulphur almost quantitatively as lead sulphide when heated with an alkali-plumbite reagent, and that boiling concentrated nitric acid rapidly oxidises cystine sulphur to sulphate and only very slowly oxidises methionine sulphur beyond sulphoxide, sulphone or, perhaps, methanesulphonic acid.

Now, Baernstein concluded, after subjecting his newer methods to tests on free cystine and methionine with the addition of some other amino acids and a little glucose, that they could be expected to yield very reliable values for the cystine and methionine contents of proteins. He was of the opinion that the sulphur of reasonably pure proteins is contained in cystine (and possibly cysteine), methionine and sulphate, and in no other forms. Generally, the methionine estimated by his "volatile-iodide" procedure checked very well with the value obtained by titration; and there can be little doubt that the titration method of estimation is the most specific for methionine that has yet been evolved. The titration method for cystine estimation cannot lay claim to great specificity, but here again the values obtained with pure proteins checked very well with published values obtained by applying more specific methods to hydrochloric acid hydrolysates.

On the other hand, Blumenthal and Clarke, having established by tests with mixtures of cystine and methionine the most satisfactory conditions for discriminating between them by nitric acid oxidation, oxidised proteins under those same conditions, and compared their results with published cystine contents as determined by the more specific methods and with the methionine contents as determined by Baernstein's "volatile-iodide" procedure. Together with data obtained by the "labile-sulphur" procedure (reaction with alkali-plumbite), and from the oxidation of proteins with bromine-water (free cystine yielding only very small amounts of sulphate), these comparisons were believed by them to indicate the existence in several proteins of small amounts of one, if not two, other new sulphur-containing amino acids. It is impossible at present categorically to deny these claims, but it will be seen that they are based upon a recurrent assumption that the cystine and methionine residues in a protein behave towards the reagents precisely as do the free amino acids, and hence the data cannot provide anything like rigorous proof either way. Indications of difference in behaviour are not lacking. Lee (133) has pointed out that the cystine contents of hydrochloric acid hydrolysates of casein are considerably lower than the disulphide contents. Bailey (7, 8) has encountered other instances and has given his opinion that, in some proteins, the cystine may be easily deaminated and decarboxylated during the hydrolytic cleavage of peptide linkages. The appearance of sulphhydryl groups in some (but not all) proteins which have been treated with acid, and the bearing of this upon cystine estimations, have been discussed by him in some detail.

Before applying the newer Baernstein and the Blumenthal and Clarke procedures to the problem of the sulphur distributions in the impure leaf proteins, it was necessary to subject them to additional tests. In particular, the effects of adding carbohydrate to at least one "pure" protein had to be examined. It was found desirable to depart from the recommended procedures in some respects. Briefly, the findings were as follows:

Baernstein's (6) procedure is to boil for eight hours in a nitrogen atmosphere about 0.5 g. of the protein with 10 ml. of concentrated hydriodic acid which has first been warmed long enough with 0.1 g. of crystalline potassium hydrogen hypophosphite to reduce any free iodine. It is an improvement to heat the acid with the minimum amount of hypophosphite necessary to give a straw-yellow coloration, to commence digestion of the proteins with an additional 0.02 g. of hypophosphite, and to continue the addition in 0.02 g. portions at intervals of not less than two minutes so long as iodine is liberated in more than barely detectable quantities. The digests of impure preparations tend to be dark from causes other than the presence of iodine, and care must be exercised to avoid overaddition of hypophosphite. Baernstein's titration of the reduced cystine finishes as a back titration of excess iodine with thiosulphate, but better end-points were obtained by changing it to a back titration of excess thiosulphate with the iodate. Results obtained by these variations in procedure were less erratic, the losses of cystine and methionine which occurred when arabinose was added to edestin were slightly diminished, and considerably increased values are obtained with the impure leaf proteins.* Small amounts of a tarry "humin" were produced when these impure preparations were digested or when carbohydrate was added to edestin before digestion. The losses occasioned by adding 20 per cent by weight of arabinose to edestin were only

* The cystine and methionine contents of preparations quoted in tables 47, 49, 53, 56 and 57 were estimated in this way.

some 7 per cent in both the cystine and methionine, and were even smaller if "Collier" pectin was substituted for arabinose. As iodides affect the Sullivan reaction for cysteine or cystine, the standards for colorimetric comparison in the methods of estimation based upon it must contain appropriate amounts of iodides. Lugg's (143) procedure was used, and in examining the hydriodic acid digests of proteins, aliquots of a gelatin digest were included in the cystine standards. Tests with free cystine and methionine showed that whereas only about 2 per cent of the cystine and 5 per cent of the methionine were lost to titration, some 6 per cent of the cystine was lost to the colorimetric estimation.

The "labile sulphur" procedure of Blumenthal and Clarke (19) cannot satisfactorily be applied to the impure leaf proteins, because the preliminary reduction cannot be carried out without involving the cystine in condensation reactions. The nitric acid oxidation procedure, however, is readily applicable, but the results obtained were a little erratic. They were much more uniform when edestin was used in the tests, and there was no detectable interference from 20 per cent by weight of arabinose added to the edestin. The arabinose was found also to be without effect upon the estimations of sulphate-sulphur and total-sulphur. From the estimations of sulphate sulphur (a) after prolonged hydrolysis, (b) after oxidising with nitric acid under suitable conditions, and (c) after completely oxidising the material (peroxide fusion), the cystine and methionine contents may be calculated on assumptions which involve the supposition that cystine (and/or cysteine) and methionine are the only sulphur-containing amino-acids entering into the composition of the protein (see Lugg, 146a). The sum of the cystine and methionine sulphurs is obtained from the difference between (a) and (c)* and is an estimation of greater precision than the individual cystine and methionine estimations calculated therefrom with the aid of (b).

Of the various means whereby information concerning the sulphur distribution may be obtained, these last, the least specific, alone show no evidence of interference by carbohydrate. All give fairly concordant estimates of the cystine and methionine contents of crystallised edestin and of many other proteins, and distinctly discordant results with the impure leaf proteins. These

* Values quoted under cystine-N plus methionine-N in tables 54 and 55 were obtained in this way.

last two procedures permit no discrimination between cysteine and cystine residues in the proteins, and there is no evidence as to whether such a partition does or does not exist in the leaf-protein preparations. In general, the cystine contents of the leaf-protein preparations estimated with the help of methods involving a preliminary acid hydrolysis are from 10 per cent to 60 per cent of the values found by titration after digestion with hydriodic acid, and these in turn represent from 70 per cent to 95 per cent of the sulphur oxidised to sulphate by nitric acid. The methionine contents estimated by titration after digestion with hydriodic acid are generally anything between 70 per cent and 100 per cent of what might be calculated from the sulphur not so oxidised. Certain correlations emerge, too. From the nitrogen and ash percentages of the impure preparations and a rough estimate of the nitrogen percentages which would be found if the preparations were pure, the amounts of organic impurity in the various preparations may be calculated approximately. For preparations made from the leaves of a particular botanical species, the disparities are normally more pronounced the greater the amount of organic impurity, and for like amounts of organic impurity, the disparities found with preparations from one species of leaf may be much greater than with preparations from another species.

The only satisfactory interpretation is that the organic impurities ordinarily present in the impure leaf-protein preparations are responsible for the disparities, in much the same way as arabinose would be responsible for similar disparities if added to edestin. That the real cystine contents are at least as great as the values obtained by titration of hydriodic acid digests is strongly suggested by the fact that, when specific colorimetric methods are applied to the same digests, reasonable confirmations of the titration values are obtained.

Estimations of Tyrosine and Tryptophan.

These amino acids have long been known to suffer modification when heated in acid solution with carbohydrates, as judged by the fact that varying amounts of the nitrogen they contain can be found in the insoluble humin (Gortner and Blish, 90; Roxas, 254). The hydrochloric acid hydrolysates of a few impure leaf-protein preparations have been found to contain virtually no tryptophan and only some 70 per cent of the tyrosine that is known to be present in the preparations. It has been shown, however, that carbohydrates do not interfere with the hydrolytic cleavage of either amino acid from edestin when alkali is used. That is to say, the presence of carbohydrate does not alter the amounts of tyrosine and tryptophan found in an alkali hydrolysate (Lugg, 145). And as far as one may judge from the indirect evidence, the hydrolytic cleavage follows a normal course and can easily be made complete. Traces of other phenols and indoles in the leaf-protein preparations are sometimes found and may prove troublesome—some extraneous indoles cannot be removed readily.

Humin Formation during Acid Hydrolysis.

The simple proteins, purified by re-crystallisation and other means, will normally yield a solution, fairly pale in colour and almost devoid of anything in the nature of a precipitate, when boiled with 5N hydrochloric or 8N sulphuric acid. Most protein preparations, however, yield a more or less dark-brown solution, and a finely divided dark precipitate. In the rather unhappy terminology adopted to describe these effects, the insoluble material is called "acid-insoluble humin," and the dark colour of the solution is attributed to an "acid-soluble humin" which may be rendered largely insoluble by partially neutralising the solution. The same terminology is used in describing the materials, of distinctly similar physical properties, produced when various carbohydrates are heated with acid. Inasmuch as the former products are known to contain nitrogen and the latter not to do so, the terms "nitrogenous-" and "non-nitrogenous humin" have been used to describe them. "Humins" have been defined by some as "substances of a melanin-like nature," and as the melanins (themselves not strictly definable) contain nitrogen, objection has sometimes been raised to use of the word "humin" in describing the materials of purely carbohydrate origin. The word "humin" will be used very loosely in the following discussion.

Whilst it is uncertain whether proteins entirely freed from carbohydrate, carbohydrate residues, uronic acids, etc., would yield "humin substances" within a normal hydrolysis period, the investigations of Roxas (254) in 1916 suggested that protein humin possibly had its origin in a carbohydrate impurity. Roxas obtained nitrogenous humins by heating pentose and hexose with various amino acids in hydrochloric acid solution. Gortner and

Blish (90) in 1915 had already found that the insoluble humin formed by heating tryptophan with carbohydrate in acid solution contained most of the tryptophan nitrogen, and Roxas was able to confirm this and to demonstrate smaller losses with tyrosine, cystine and the basic amino acids. He found that some sugars were more reactive than others and concluded that the formation of furfural and substituted furfurals from the sugars may have been largely responsible for the subsequent appearance of humin. Both he and Gortner (89) obtained evidence of pronounced activity between furfural and some amino acids in acid solution, with the formation of nitrogenous humins.

Subsequent work by others has extended these findings, and it has thus come to be believed that humin formation is due to carbohydrate inclusion and to a greater or lesser extent may take place through the condensation of furfuraldehydes with themselves and with amino acids. Various condensation mechanisms have been proposed. Apart from an ill-defined adsorption process and a generalised aldehyde-amine condensation (see, for example, Levy, 136) which may be operative in the case of many amino acids, more specific condensations must be considered, as in the case of cysteine, tyrosine and tryptophan, for mercaptans readily condense with aldehydes to form mercaptals, phenols condense with aldehydes forming substances akin to "bakelite," and indoles readily form coloured condensation products with aldehydes (a type of reaction which has been exploited in methods of estimating tryptophan). These three amino acids, incidentally, are among the most susceptible to loss associated with humin formation. Nor can the possibility be ignored that some amino acids, not directly able to contribute much to the formation of nitrogenous humin, may yet do so indirectly and at least temporarily in consequence of their peptide linkages with amino acids which do condense readily. The presumption is that they would be freed by continued hydrolysis, but their own reactivity in condensation reactions may be increased or their peptide linkages protected in some way. Bergmann (11), for example, considered that peptide groups may react with aldehyde. Roxas (254) believed that the presence of one amino acid could increase the susceptibility of another to enter into humin formation.

Miller and Chibnall (171), having suspected from the large acid-humin formation the presence of considerable carbohydrate impurity in their leaf-protein preparations, were disappointed by their inability to isolate more than several per cent by weight of a polysaccharide after alkali-hydrolysis and to obtain more than some 1 per cent by weight of furfural by distilling the preparations with hydrochloric acid under the Tollens conditions. But we know now that the yield of furfural from pentoses and pentosans under the Tollens conditions is very greatly reduced by the presence of protein (Lee, private communication), and the low yields obtained by Miller and Chibnall can be taken as an indication, not that the amounts of substance capable of yielding furfural were correspondingly small, but that they must have been very much larger.

It is of some interest to enquire into the composition of the humin obtained from a few sources. In the first place, we may consider the humin obtained by Bailey on heating 1 g. of edestin in acid solution with 0.1 g. of arabinose. The weight of humin varied, with a mean of 0.082 g., and contained 0.0048 g. of nitrogen and 0.0010 g. of sulphur. Like the humins described by Roxas, these were obtained by adjusting the acidity of the hydrolysis mixture to the region of minimum solubility, which appears to lie generally at about pH 2.5. We may reasonably assume that the humin sulphur had its origin in cystine and methionine, and that each atom of sulphur would still have an atom of nitrogen associated with it, and so 0.0045 g. of the humin nitrogen must have come from other sources. We are assuming here that all peptide linkages were hydrolysed, and in fact it is only upon this basis that the speculation can be continued. If we assume from the work of Roxas (254) that some 70 per cent of the tryptophan nitrogen and 15 per cent of the tyrosine nitrogen entered this humin, we can account for a further 0.0019 g. of the nitrogen. The remaining nitrogen, 0.0026 g., represents only 1.4 per cent of the original protein nitrogen and its origin is probably to be sought chiefly among arginine, histidine, proline and hydroxyproline, which together represent a fairly large fraction of the protein nitrogen. One of these amino acids may have contributed overwhelmingly to the humin nitrogen and the others negligibly; or each may have contributed very little, approximately in the proportion in which it occurs in the protein.

To ignore the part (if any) played by the furfural double-bonds, the condensation reactions presumably involve elimination of water between furfural molecules and between furfural and amino acid molecules, and, in the latter case, it might be anticipated that an amino acid molecule normally loses only one or two hydrogen atoms in the reaction. On this basis, the contribution of the amino acids to the weight of the humin may be calculated roughly. By taking 120 as the mean residue weight of the cystine, methionine, tryptophan and tyrosine moieties, their contribution would be 0.021 g. Assignment of an appropriate mean residue weight to the rest of the nitrogen in the humin, based upon the assumption that histidine, arginine, proline and hydroxyproline are the chief contributors, can be little better than a guess. Taking 70 as the weight associated with each nitrogen atom, the contribution to the weight of the humin would be 0.013 g. That is to say, of the 0.082 g. of humin, 0.034 g. would have been contributed by the amino acids and 0.048 g. by the furfural or arabinose. There is some basis for the assumption that the residues in a non-nitrogenous furfural humin may be regarded as dehydrated furfural molecules, but it may not be the case for all the furfural residues in a nitrogenous humin. If it is the case, however, then in the particular humin under consideration there would appear to have been two furfural residues (each of weight 78) to every amino acid residue. The assumptions concerning the nature of the condensation of furfural with itself acquire some support from experiment. A pentose should yield 64 per cent of its weight of furfural, and thus, by these assumptions, some 52 per cent of its weight of humin. Actually, a pentose yields some humin and about 51 per cent of its weight of furfural in the Tollens procedure. Gortner (89) obtained 76 parts of non-nitrogenous humin from 100 parts of furfural by heating in acid solution, thus suggesting a yield of 49 parts of humin from 100 of pentose in place of the calculated 52 parts. Of the 0.052 g. of furfural residues which could have appeared (on this basis) in the humin obtained by Bailey, most (0.048 g.) has already been accounted for, and it would therefore seem that there could have been very little soluble humin present. Cystine and methionine are both known to have contributed to the soluble fraction, and the rest of the tryptophan and some of the tyrosine must have been present, too, but the total contribution by all the amino acids involved must have been small. It must have been some three times as large if, despite the evidence of Gortner's (89) experiment, the condensation of furfural with itself is of the aldehyde-aldol or aldehyde-paraldehyde type, but that would still be small.

These speculations have been based upon very meagre and

unprecise data and upon many insecure assumptions. That such uncertain foundations have to be used is itself an illuminating commentary upon the extent of our present knowledge of humin-formation.

In the second place, we may consider the humin obtained from two cocksfoot leaf-protein preparations which contained very little ash. The percentages of nitrogen in these preparations were 13.05 and 14.1, and the weights of humin as percentages of the weights of the corresponding preparations were 19.3 and 15.0, the humins containing 7.2 and 5.6 per cent of the protein nitrogen, respectively. On the basis of the previous calculations, about 34 per cent and 37 per cent, respectively, of weights of these humins could be attributed to the amino acid residues involved, and, if we assume that the impurities in the preparations were all of a pentosan character, we find that the percentages of nitrogen in the "impurity-free" materials were 16.7 and 16.8, respectively. The close agreement may be fortuitous, but it is favoured by the fact that the known differences in the amino acid compositions of the preparations are very small. The order of magnitude, too, is supported by other evidence. Miller (170), from an approximately determined amino acid composition of several preparations, concluded that the "pure" proteins would contain about 16 per cent of nitrogen. Subsequent and more precise analytical work with preparations from cocksfoot leaves leads to the conclusion that the value may be nearer 16.5 per cent in these cases. Chibnall and Grover (43) have quoted values of 16.25 per cent and 16.7 per cent, respectively, for actual (impure) preparations from spinach and broad-bean leaves.

Of the real nature of the common impurity in the leaf-protein preparations, very little is definitely known. In association with the protein, it forms fairly clear solutions both on the acid and alkaline sides of the apparent isoelectric region of the protein, and it is precipitated with the protein on flocculation. In amount it may vary widely, so presumably only a fraction of it, if any, can be prosthetically associated and escape the category of "genuine impurity." From the solubility relationships it seems rather unlikely that it (or most of it) can be a simple pentosan. It may be a pectin, but the most attractive hypothesis is that it belongs to the mucilages; for all mucilages are not of the same composition, and so we are left with a tentative explanation of the very variable cystine losses resulting from acid-digestion of preparations which are of similar degrees of "purity," but of different origins among the species of plant leaves. The point cannot be stressed, however, because the proteins themselves are not of identical composition.

APPENDIX II

The Estimation of Nitrogenous Bases, Dicarboxylic Acids and Amide Nitrogen in the Impure Leaf-Protein Preparations.

IN Appendix I, Dr. Lugg has pointed out the difficulties which attend the estimation, after acid hydrolysis, of cystine, methionine, tryptophan and tyrosine in the leaf proteins made by methods described in Chapters VI and VII, due to impurities—he suggests that they may be mucilages—which accompany the proteins in the leaves and cannot be separated from them during the course of preparation. These findings have necessitated a careful re-examination of the methods to be used for determining certain other amino acids and also the amide nitrogen, for we were anxious, if possible, to estimate these not only in the impure leaf-protein preparations, but also in products prepared from dried leaves by methods described below, which contain the whole protein of the leaf.

The Estimation of Nitrogenous Bases by the Method of Block (18).

The following is a résumé of Dr. Tristram's results, which will be published in full later.

A. *Losses due to the presence of pentosan impurities.* Solutions containing such amounts of arginine, histidine and lysine as would be given by 2.5 g. of a leaf protein (N = 17 per cent) were added to the requisite amount of sulphuric acid and the mixture, after hydrolysis with or without added impurities, was passed through the Block procedure. The recoveries of arginine, histidine and lysine (after recrystallisation of the picrate) are given in table 95 the arginine value carrying a correction for arginine silver of 1.1 mg. of arginine nitrogen per 100 ml. of solution (Gulewitsch, 105).

The arginine loss in the presence of 20 per cent arabinose was confirmed by a parallel experiment with edestin; in the absence of the impurity the arginine content found was 15.2 per cent,

TABLE 95.
(From Tristram, 325.)

Impurity added	Arginine recovered	Histidine recovered	Lysine recovered
g.	%	%	%
None	94.2	98	93
Arabinose, 0.25	86.5	97	95
Arabinose, 0.5	83.5	100	93
Hemicellulose, 2.5	73.0	68	90

and in its presence 13.05 per cent, representing a recovery of 85 per cent in the latter case.

B. *Estimation of bases in the leaf-protein preparations.* The nitrogen content of these preparations varied between 12.5 per cent and 15.5 per cent, and the amount of impurity present was calculated on the assumption that the pure protein would contain 17 per cent. In computing the values given in the various tables in Chapters VI and VII, the recoveries given in table 95 have been taken as valid, the amount of impurity present being taken as equivalent either to 10 per cent or to 20 per cent of arabinose.

C. *Estimation of bases in the "whole protein" of dried leaves.* The dried and ground leaf material was first treated at the boil with dilute citric acid solution to remove all soluble non-protein nitrogenous products, sugars, etc., and then with graded strengths of alcohol and finally with ether to remove lipoids. The almost white residue thus obtained was considered to include the whole protein of the leaf and no other (insoluble) nitrogenous product. In one particular experiment, the dried grass contained 4.2 per cent of nitrogen, and, after extraction in this way, the leaf-residue contained 5.6 per cent nitrogen. If this nitrogen represented protein then, from the point of view of hydrolysis, the impurity present was about 200 per cent, and, to reduce this, the leaf-residue was boiled for a few hours with sufficient 7.5 per cent sulphuric acid such that, on filtration and concentration of the filtrate, the correct conditions for a Block analysis could be achieved. About 92 per cent of the leaf-residue nitrogen was extracted in this way, and the final residues being almost black in colour, it was assumed that the unextracted nitrogen had passed into humin. By difference, the extracted material contained 7.9 per cent nitrogen, and the subsequent hydrolysis thus took place in the presence of about 100 per cent impurity. Reference to table

95 shows that both arginine and histidine suffer much destruction under such conditions, and lysine was regarded as the only base which could be estimated with any reliability, the recovery being taken as 90 per cent.

The Estimation of Nitrogenous Bases by the Method of Van Slyke.

Miller (169a) has called attention to the high values for histidine and lysine obtained when this method is applied to cocksfoot protein. The question will be discussed more fully in a forthcoming publication by Tristram (325); meanwhile some typical results given in table 96 show that my earlier base analyses were unreliable, and they have therefore been withdrawn (cf. p. 140).

TABLE 96.

(Figures given are in percentages of total protein-N.)

	Spinach		Zea Mays	
	Van Slyke (Chibnall, 34)	Block (Tristram, 325)	Van Slyke (Chibnall and Nolan, 47)	Block (Tristram, 325)
Arginine-N	13.80	14.1	14.69	14.4
Histidine-N	3.89	2.2	4.70	2.1
Lysine-N	9.63	6.2	8.78	6.1

The Estimation of Aspartic Acid and Glutamic Acid.

The method used was a modification of that of Foreman (82), full details of which will be published later. The hydrolysis of these two acids in the presence of 20 per cent arabinose leads to no detectable loss of amino nitrogen.

The Estimation of Amide Nitrogen.

The well-known Sachsse conditions, hydrolysis with 5 per cent hydrochloric acid for three hours, were used in all cases. The usual hydrolysis with 20 per cent hydrochloric acid for 20 hours leads to decomposition and enhanced values for amide nitrogen, as shown below.

The variations in the values for amide nitrogen as between the extracted protein and leaf-residues which I found in some of my early experiments with bean leaves (31) were therefore not valid.

TABLE 97.

(Unpublished data.)

	Total-N %	Amide-N as a percentage of total protein-N	
		After hydrolysis with 5% HCl for 3 hours	After hydrolysis with 20% HCl for 20 hours
Preparation of rough-stalked meadow grass protein	13.6	4.8	5.9
Whole protein of perennial rye-grass*	5.1	5.2	6.3
Treated residues of perennial rye-grass†	3.5	5.2	6.6

* The dried and powdered grass was first treated with citric acid, alcohol and ether, as described earlier in this Appendix.

† Leaf-residues after extraction of protein by the ether-water method, which had been subsequently treated as in above footnote.

APPENDIX III

The Lipoid Fraction of Chloroplasts.

ALTHOUGH the chemical composition of the lipoid fraction of chloroplasts is rather outside the main theme of this book, I should like to make brief mention of it here to amplify my discussion on the chloroplastic proteins given in Chapter VII.

In 1926, Channon and Chibnall investigated the ether-extract of cabbage leaf protoplasm (referred to by them as "cytoplasm"; see footnote, p. 129). The leaves were ground with water, the green juice was expressed and then warmed to flocculate the dispersed protoplasm. The separated product was extracted with ether and, since the cytoplasm undoubtedly contains but little lipoid, it follows that the ether-soluble substances thus obtained were derived in large, probably very large, part from the chloroplasts (nucleoplasm, from lack of evidence, must be neglected). Samples of cabbage leaves collected at different times of the year gave products exhibiting the fairly constant protein: lipoid* ratio of about 3.1:1 (39) and, as the ratio of cytoplasmic to chloroplastic protein is about 1:1 (43), it follows that the protein: lipoid ratio in the chloroplastic material is 3.1:2 or 61:39, in fair agreement with corresponding values for spinach (cf. table 46).

In view of the probable colloidal relationship between protein and phosphatide, the components of the latter are interesting. In the case of cabbage it consisted chiefly of calcium phosphatidate (40, 28), lecithin and kephalin being absent, but all three were present in cocksfoot (312) and runner bean (113). Other points worthy of note are the high degree of unsaturation of the glycerides, and the presence of waxes: cabbage—paraffin and ketone (29); cocksfoot—primary alcohols (221).

---

* Chibnall and Channon (41) voiced objections to the use of the term "lipoid" in the case of ether-extracts from leaves, which contain chloroplastic pigments. The term is used here for simplicity and to avoid confusion in comparing results with those of Menke.

A detailed analysis of cabbage and cocksfoot ether-soluble material (unfortunately no *ad hoc* data for spinach are available) is given in table 98.

The metabolism of phosphatidic acid in seedlings is interesting. Channon and Foster (30) found that in wheat germ it was present as a mixture of calcium, magnesium and potassium salts. In the runner bean (Jordan and Chibnall, 113), the embryo axis contained only magnesium salt. This increased in amount during germination until the development of the prophylls, when a rapid change over to the calcium salt took place, the (fat soluble) magnesium being possibly utilised in the synthesis of chlorophyll.

TABLE 98.

(Compiled from 41, 312, 221, and unpublished data.)

Ether-soluble substances, derived in large part from the chloroplasts of cabbage and cocksfoot.

(Figures given are in percentages of total ether-extract.)

Pigments	Cabbage (*Brassica oleracea*)	Cocksfoot (*Dactylis glomerata*)
Chlorophyll (α and β)	9.3	19.0
Carotene	0.5	0.9
Xanthophyll	0.8	1.75
Glyceride fatty acids	17.5*	38.0†
Unsaponifiable material, containing:		
Wax	12.3	23.3
Sterols	4.5	3.0
Undetermined	13.3	8.4
Crude phosphatide,‡ containing:		
Calcium phosphatidate	18.4	0.9
Lecithin	–	0.6
Kephalin	–	

* Iodine value 200.
† Iodine value 150.
‡ Considerable losses occur in fractionating these products.

BIBLIOGRAPHY

1. ABDERHALDEN and HERRICK 1905. *Z. physiol. Chem.* 45 : 479.
2. ADLER, DAS, v. EULER and HEYMAN 1938. *Comp. rend. Labor. Carlsberg* 22 : 15.
3. ANDREWS 1903. *Jahrb. wiss. Bot.* 38 : 1.
4. ANNAU, BANGA, GÖZSY, HUSZAK, LAKI, STRAUB and SZENT-GYÖRGYI 1935. *Z. physiol. Chem.* 236 : 1.
5. BAERNSTEIN 1934. *J. Biol. Chem.* 106 : 451.
6. —— 1936. *J. Biol. Chem.* 115 : 25.
7. BAILEY 1937. *Biochem. J.* 31 : 1396.
8. —— 1937. *Biochem. J.* 31 : 1406.
9. BAUMANN 1882. *Pflüger's Arch.* 29 : 419.
10. BENNET-CLARK 1933. *New Phytologist* 32 : 37, 128, 197.
11. BERGMANN 1923. *Collegium* 132.
12. BENTE 1874. *Jour. Landw.* 22 : 113.
13. BEYER 1867. *Arch. d. Pharm.* 201.
14. —— 1867. *Land. Vers. Sta.* 9 : 186.
15. BISHOP 1930. *J. Inst. Brewing* 27 (N.S.) : 323.
16. BJÖRKSTÉN 1930. *Biochem. Z.* 225 : 1.
18. BLOCK 1934. *J. Biol. Chem.* 106 : 457.
19. BLUMENTHAL and CLARKE 1935. *J. Biol. Chem.* 110 : 343.
20. BORODIN 1876. *Bot. Jahresber.* 4 : 919.
21. —— 1878. *Bot. Zeitung.* 36 : 802.
22. BOUSSINGAULT 1868. *Agronomie, chemie agr. et physiol.* 4 : 245 ff.
23. BRAUNSTEIN and KRITSMAN 1937. *Enzymologia* 2 : 129.
24. —— and KRITSMAN 1937. *Nature* 140 : 503.
25. BURKHART 1938. *Plant Physiol.* 13 : 265.
26. BUTKEWITSCH 1909. *Biochem. Z.* 16 : 411.
27. —— 1902. Tageblatt des XI Naturforscherkongresses in St. Petersburg.
28. CHANNON and CHIBNALL 1927. *Biochem. J.* 21 : 1112.
29. —— and CHIBNALL 1929. *Biochem. J.* 23 : 168.
30. —— and FOSTER 1934. *Biochem. J.* 28 : 853.
31. CHIBNALL 1922. *Biochem. J.* 16 : 343.
32. —— 1923. *Ann. Bot.* 37 : 511.
33. —— 1923. *J. Biol. Chem.* 55 : 333.

34. —— 1924. *J. Biol. Chem.* 61:303.
35. —— 1924. *Biochem. J.* 18:387.
37. —— 1924. *Biochem. J.* 18:395.
38. —— 1926. *J. Am. Chem. Soc.* 48:728.
39. —— and CHANNON 1927. *Biochem. J.* 21:225.
40. —— and CHANNON 1927. *Biochem. J.* 21:233.
41. —— and CHANNON 1929. *Biochem. J.* 23:176.
42. —— and GROVER 1926. *Ann. Bot.* 40:491.
43. —— and GROVER 1926. *Biochem. J.* 20:108.
45. ——, MILLER, HALL and WESTALL 1933. *Biochem. J.* 27:1879.
46. —— and NOLAN 1924. *J. Biol. Chem.* 62:173.
47. —— and NOLAN 1924. *J. Biol. Chem.* 62:179.
49. —— and SCHRYVER 1920. *J. Physiol.* 54: *Proc.* July.
50. —— and SCHRYVER 1921. *Biochem. J.* 15:60.
51. —— and SAHAI 1931. *Ann. Bot.* 45:489.
52. —— and WESTALL 1932. *Biochem. J.* 26:122.
53. CLAUSEN 1890. *Land. Jahrb.* 19:914.
54. CLIFT and COOK 1932. *Biochem. J.* 26:1788.
55. COHN 1935. *Ann. Review Biochem.* 4:93.
56. CORBET 1935. *Biochem. J.* 29:1086.
57. COSSA 1875. *Gaz. chim. ital.* 5:314.
58. CRAMPTON and FINLAYSON 1935. *Empire J. Exp. Agric.* 3:331.
59. DAMODARAN 1932. *Biochem. J.* 26:235.
60. ——, JAABACK and CHIBNALL 1932. *Biochem. J.* 26:1704.
61. —— and NAIR 1938. *Biochem. J.* 32:1064.
62. DELEANO 1912. *Z. physiol. Chem.* 80:79.
63. —— 1912. *Jahrb. wiss. Bot.* 51:541.
64. DENNY 1932. *Contrib. Boyce Thompson Inst.* 4:65.
65. DESSAIGNES and CHAUTARD 1848. *Jour. pharm. chim.* (3) 13:246.
66. DETMER 1879–1881. *Jahrb. wiss. Bot.* 12:236.
67. DIXON and ATKINS 1913. *Sci. Proc. Royal Dublin Soc.* 13 (N.S.):422.
68. —— and BALL 1924. *Sci. Proc. Royal Dublin Soc.* 17 (N.S.):263.
69. DUMAS and CAHOURS 1842. *Ann. chim. phys.* (3), 6:385.
70. ECKERSON 1924. *Bot. Gaz.* 77:377.
71. EDLBACHER 1926. *Z. physiol. Chem.* 157:106.
72. —— and KRAUS 1930. *Z. physiol. Chem.* 191:225.
73. EGGLETON 1935. *Biochem. J.* 29:1389.
74. EICHHORN 1867. *Land. Vers. Sta.* 9:275.

75. ELLIOTT, BENOY and BAKER 1935. *Biochem. J.* 29 : 1937.
76. EMMERLING 1880. *Land. Vers. Sta.* 24 : 113.
77. —— 1887. *Land. Vers. Sta.* 34 : 1.
78. —— 1900. *Land. Vers. Sta.* 54 : 215.
79. ENDRES 1935. *Ann.* 517 : 109.
80. v. EULER, ADLER, GÜNTHER and DAS 1938. *Z. physiol. Chem.* 254 : 61.
80a. FAGAN and ASHTON 1938. *Welsh J. Agric.* 14 : 160.
81. FISCHER 1936. *Z. Bot.* 30 : 449.
82. FOREMAN 1914. *Biochem. J.* 8 : 463.
83. —— 1938. *J. Agric. Sci.* 28 : 135.
84. FOURCROY 1789. *Ann. chim.* (1), 3 : 252.
85. FREY-WYSSLING 1937. *Protoplasma* 29 : 279.
86. GARREAU 1851. *Ann. sci. nat.* (3 Bot.) 15 : 1.
87. GODLEWSKI 1903. *Bull. intern. acad. sci. Cracovie* (B) 313.
88. —— 1911. *Bull. intern. acad. sci. Cracovie* (B) 623.
89. GORTNER 1916. *J. Biol. Chem.* 26 : 177.
90. —— and BLISH 1915. *J. Amer. Chem. Soc.* 37 : 1630.
91. GORUP-BESANEZ 1874. *Ber.* 7 : 146.
92. —— 1874. *Ber.* 7 : 569.
93. —— 1874. *Ber.* 7 : 1478.
94. —— 1877. *Ber.* 10 : 780.
95. GOUWENTAK 1929. *Rec. Trav. Bot. Neerl.* 26 : 19.
96. GREEN, R. 1887. *Phil. Trans. Royal Soc. Lond.* B 178 : 39.
97. —— 1890. *Proc. Royal Soc. Lond.* B 48 : 370.
98. GREEN, D. E. 1936. *Biochem. J.* 30 : 2095.
99. GREENHILL and CHIBNALL 1934. *Biochem. J.* 28 : 1422.
100. GREGORY 1937. *Ann. Review Biochem.* 6 : 557.
101. —— and BAPTISTE 1936. *Ann. Bot.* 50 : 579.
102. —— and SEN 1937. *Ann. Bot.* 1 (N.S.) : 521.
104. GROVER and CHIBNALL 1928. *Biochem. J.* 21 : 857.
105. GULEWITSCH 1899. *Z. physiol. Chem.* 27 : 178.
106. HANSTEEN 1899. *Jahrb. wiss. Bot.* 34 : 417.
107. HARTIG 1855. *Bot. Zeitung.* 13 : 881; 1856. 14 : 257.
108. —— 1858. *Entwickelungsgeschichte des Pflanzenkeims.* Leipzig.
109. HLASIWETZ and HABERMANN 1871. *Ann.* 159 : 304.
110. —— and HABERMANN 1873. *Ann.* 169 : 150.
111. HOPPE-SEYLER 1870. *Handbuch der physiologisch- u. pathologisch-chem. Analyse.* 3 Aufl. p. 208.
112. HOSAEUS 1868. *Arch. d. Pharm.* 135 : 42.
113. JORDAN and CHIBNALL 1933. *Ann. Bot.* 47 : 163.

BIBLIOGRAPHY 289

114. KARRER, SMIRNOFF, EHRENSPERGER, VAN SLOOTEN and KELLER 1924. *Z. physiol. Chem.* 135 : 129.
115. KASSELL and BRAND 1938. *J. Biol. Chem.* 125 : 145.
116. KELLNER 1879. *Land. Jahrb.* 8 : supplement 1, 243.
117. KIESEL 1906. *Z. physiol. Chem.* 49 : 72.
118. ——, BELOZERSKY, AGOTOW, BIWSCHICH and PAWLOWA 1934. *Z. physiol. Chem.* 226 : 73.
118a. KLEIN and TAUBÖCK 1932. *Biochem. Z.* 251 : 10.
118b. KNOOP and MARTIUS 1936. *Z. physiol. Chem.* 242 : I.
119. KNOOP and OESTERLIN 1925. *Z. physiol. Chem.* 148 : 294.
120. KNOP and WOLF 1868. *Land. Vers. Sta.* 10 : 13.
121. KOLBE 1862. *Ann.* 121 : 232.
122. KRAUCH 1882. *Land. Vers. Sta.* 27 : 383.
123. KREBS 1933. *Z. physiol. Chem.* 218 : 157.
124. —— 1933. *Z. physiol. Chem.* 217 : 191.
125. —— 1935. *Biochem. J.* 29 : 1951.
126. —— 1937. *Lancet.* September 25, p. 736.
127. —— and JOHNSON 1937. *Enzymologia* 4 : 148.
128. KÜHNE 1876. *Verhandl. Naturhist.-med. Vereins zu Heidelberg* 1 : 236.
129. KULTZSCHER 1932. *Planta* 17 : 699.
130. KÜSTER 1935. *Die Pflanzenzelle.* Jena.
131. KUTSCHER 1901. *Z. physiol. Chem.* 32 : 413.
132. LASKOWSKY 1874. *Land. Vers. Sta.* 17 : 219.
133. LEE 1935. *Australian J. Exp. Biol. Med. Sci.* 13 : 229.
134. LEMOIGNE, MONGUILLON and DESVEAUX 1937. *Bull. soc. chim. biol.* 19 : 671.
135. ——, MONGUILLON and DESVEAUX 1937. *Compt. rend.* 204 : 1841.
136. LEVY 1933. *J. Biol. Chem.* 99 : 767.
137. LEPESCHKIN 1936. *Biodynamica* 19 : 1.
138. LIEBALDT 1913. *Z. Bot.* 5 : 65.
139. Löw 1875. *Pflüger's Arch.* 22 : 503.
140. —— 1892. Cited by Schulze (274, p. 121) from a reprint of unknown origin.
141. —— 1885. *J. pr. Chem.* 31 (N.F.) : 129.
142. LUGG 1932. *Biochem. J.* 26 : 2160.
143. —— 1933. *Biochem. J.* 27 : 668.
144. —— 1933. *Biochem. J.* 27 : 1022.
145. —— 1938. *Biochem. J.* 32 : 775.
146a. —— 1938. *Biochem. J.* 32 : 2114.
146b. —— 1938. *Biochem. J.* 32 : 2123.

146c. —— 1939. *Biochem. J.* 33.
147. MacDougal 1920. *Carnegie Inst. Wash. Publ.* No. 297.
148. —— and Spoehr 1920. *Proc. Amer. Phil. Soc.* 59 : 150.
149. Marston 1937. *Fourth International Grassland Congress. Report*: 26.
150. Martius 1937. *Z. physiol. Chem.* 247 : 104.
152. Martius and Knoop 1937. *Z. physiol. Chem.* 246 : I.
153. Maskell and Mason 1929. *Ann. Bot.* 43 : 615.
154. —— and Mason 1930. *Ann. Bot.* 44 : 657.
155. Mason and Maskell 1929. *Ann. Bot.* 42 : 189.
158. McKee, H. S. 1937. *New Phytologist* 36 : 33, 240.
159. McKee, M. C. and Lobb 1938. *Plant Physiol.* 13 : 407.
160. Menke 1934. *Protoplasma* 21 : 279.
161. —— 1934. *Protoplasma* 22 : 56.
162. —— 1938. *Z. Bot.* 32 : 273.
163. Mercadante 1875. *Gaz. chim. ital.* 5 : 187.
164. Merlis 1897. *Land. Vers. Sta.* 48 : 419.
165. Mayer 1870. *Lehrbuch der Agrikulturchemie*, p. 154.
166. Meyer 1918. *Flora* 111–112 : 85.
167. —— 1918. *Ber. bot. Ges.* 36 : 235.
168. —— 1920. *Morphologische und physiologische Analyse der Zelle der Pflanzen und Tiere.* Jena.
169. Michael 1935. *Z. Bot.* 29 : 385.
169a. Miller 1935. *Biochem. J.* 29 : 2344.
170. —— 1936. *Biochem. J.* 30 : 273.
171. —— and Chibnall 1932. *Biochem. J.* 26 : 392.
172. Molisch 1916. *Z. Bot.* 8 : 124.
173. Morris and Wright 1933. *J. Dairy Res.* 4 : 177.
174. —— and Wright 1933. *J. Dairy Res.* 5 : 1.
175. ——, Wright and Fowler 1936. *J. Dairy Res.* 7 : 97.
176. Mothes 1925. *Planta* 1 : 472.
177. —— 1929. *Planta* 7 : 585.
178. —— 1931. *Planta* 12 : 686.
179. —— 1933. *Planta* 19 : 117.
180. —— 1933. *Ber. bot. Ges.* 51 : (31).
181. Müller 1887. *Land. Vers. Sta.* 33 : 311.
182. Nasse 1872. *Pflüger's Arch.* 6 : 582.
183. —— 1873. *Pflüger's Arch.* 7 : 139.
184. —— 1874. *Pflüger's Arch.* 8 : 381.
185. Nightingale 1937. *Bot. Reviews* 3 : 85.
186. ——, Schermerhorn and Robbins 1932. *Plant physiol.* 7 : 565.

BIBLIOGRAPHY

187. NOACK 1927. *Biochem. Z.* 183:135.
188. NUCCORINI 1930. *Ann. chim. applicata.* 20:239.
189. OSBORNE 1924. *The Vegetable Proteins.* London.
190. —— and CAMPBELL 1897. *J. Am. Chem. Soc.* 19:454.
191. —— and CAMPBELL 1898. *J. Am. Chem. Soc.* 20:348.
192. —— and CLAPP 1907. *J. Biol. Chem.* 3:219.
193. —— and CLAPP 1907. *Am. J. Physiol.* 18:295.
194. —— and GILBERT 1906. *Amer. J. Physiol.* 15:333.
195. —— and HEYL 1908. *J. Biol. Chem.* 5:187, 197.
198. —— and WAKEMAN 1920. *J. Biol. Chem.* 42:1.
199. ——, WAKEMAN and LEAVENWORTH 1921. *J. Biol. Chem.* 49:63.
200. PAECH 1935. *Planta* 24:78.
201. PALLADIN 1888. *Ber. bot. Ges.* 6:205, 296.
202. PASTEUR 1851. *Ann. chim. phys.* (3), 31:72.
203. —— 1852. *Ann.* 82:324.
204. PEARSALL and BILLIMORIA 1936. *Nature* 138:801.
205. —— and BILLIMORIA 1937. *Biochem. J.* 31:1743.
206. —— and BILLIMORIA 1938. *Ann. Bot.* 2 (N.S.):317.
207. PELOUZE 1833. *Ann.* 5:283.
208. PFEFFER 1872. *Jahrb. wiss. Botan.* 8:429.
209. —— 1873. *Monatsheft. d. Berliner Ak.* p. 780.
210. —— 1876. *Land. Jahrb.* 5:87.
211. —— 1880. *Pflanzenphysiologie* 1:300.
212. —— 1897. *Plant Physiology.* 2d Ed. Translated by EWART.
213. PFLÜGER 1875. *Pflüger's Arch.* 10.
214. PFENNINGER 1909. *Ber. bot. Ges.* 27:227.
215. PHILLIS and MASON 1937. *Nature* 140:370.
216. PIRIA 1844. *Compt. rend.* 19:575.
217. —— 1848. *Ann. chim. phys.* (3) 22:160.
218. PIUTTI 1887. *Gaz. chim. ital.* 17:519; 1888, 18:457.
219. PLISSON 1828. *J. Pharm.* 14:177.
220. POLLARD and CHIBNALL 1934. *Biochem. J.* 28:326.
221. ——, CHIBNALL and PIPER 1931. *Biochem. J.* 25:2111.
222. PRIANISCHNIKOW 1895. *Land. Vers. Sta.* 45:247.
223. —— 1895. *Land. Vers. Sta.* 46:459.
224. —— 1899. *Ber. bot. Ges.* 17:151.
225. —— 1899. *Land. Vers. Sta.* 52:137.
226. —— 1899. *Land. Vers. Sta.* 52:347.
227. —— 1900. *Ber. bot. Ges.* 18:285.
228. —— 1904. *Ber. bot. Ges.* 22:35.
229. —— 1922. *Ber. bot. Ges.* 40:242.

230. —— 1924. *Revue gén. de Bot.* 36 : 108.
231. —— and SCHULOW 1910. *Ber. bot. Ges.* 28 : 253.
232. PRUNTY 1933. *Biochem. J.* 27 : 387.
233. PUCHER, VICKERY and LEAVENWORTH 1934. *Indus. and Engin. Chem. Anal. Ed.* 6 : 190.
235. ——, VICKERY and WAKEMAN 1934. *Indus. and Engin. Chem. Anal. Ed.* 6 : 140.
236. ——, VICKERY and WAKEMAN 1934. *Indus. and Engin. Chem. Anal. Ed.* 6 : 288.
236a. QUASTEL and WOOLF 1926. *Biochem. J.* 20 : 545.
237. RAHN 1932. *Planta* 18 : 1.
238. RICHARDS 1932. *Ann. Bot.* 46 : 367.
239. —— 1938. *Ann. Bot.* 2 (N.S.) : 491.
240. —— and TEMPLEMAN 1936. *Ann. Bot.* 50 : 367.
241. RICHARDSON 1934. *Proc. Royal Soc. Lond.* B 115 : 142.
242. —— 1934. *Proc. Royal Soc. Lond.* B 115 : 170.
243. RITTHAUSEN 1860 onwards. Numerous references in Osborne (189).
244. —— 1866. *J. pr. Chem.* 99 : 454.
245. —— 1868. *J. pr. Chem.* 103 : 233.
246. —— 1868. *J. pr. Chem.* 103 : 65, 193, 273.
247. —— 1869. *J. pr. Chem.* 107 : 218.
248. —— 1872. *Die Eiweisskörper.* Bonn.
249. —— and KREUSLER 1871. *J. pr. Chem.* (II), 3 : 314.
250. ROBERTSON 1928. *Council Sci. Ind. Res. Australia,* Bull. 39.
251. ROBINSON 1929. *New Phytologist* 28 : 117.
252. ROSE 1938. *Physiol. Reviews* 18 : 109.
253. ROUELLE 1773. *Journal de médicine, chirurgie, pharmacie, etc.* 39 : 250; 40 : 59.
254. ROXAS 1916. *J. Biol. Chem.* 27 : 71.
255. RUHLAND and WETZEL 1926. *Planta* 1 : 558.
256. —— and WETZEL 1927. *Planta* 3 : 765.
257. —— and WETZEL 1929. *Planta* 7 : 503.
258. SACHS 1882. *Experimentalphysiologie,* p. 380.
259. —— 1862. *Flora* 45 : 129.
260. SACHSSE 1873. *Land. Vers. Sta.* 16 : 61.
261. —— 1876. *Sitz. d. Naturforsch. Ges. Leipzig* 3 : 26.
262. —— and KORMANN 1874. *Land. Vers. Sta.* 17 : 321.
263. SAID 1934. Thesis, Imperial College of Science, London.
264. SALOMON 1878. *Ber.* 11 : 574.
265. SAPOSCHNIKOW 1895. *Bot. Zent.* 63 : 246.
266. SCHULTZE 1861. *Arch. Anat. u. Physiol.* 1.

267. Schulze, E. 1877. *Land. Vers. Sta.* 20:117.
268. —— 1878. *Land. Jahrb.* 7:411.
269. —— 1879. *Bot. Zeitung.* 37:210.
270. —— 1880. *Land. Jahrb.* 9:689.
271. —— 1885. *Land. Jahrb.* 14:713.
272. —— 1888. *Land. Jahrb.* 17:683.
273. —— 1891. *Ber.* 24:1098.
274. —— 1892. *Land. Jahrb.* 21:105.
275. —— 1893. *Z. physiol. Chem.* 17:193.
276. —— 1895. *Z. physiol. Chem.* 20:327.
277. —— 1895–1896. *Z. physiol. Chem.* 21:392.
278. —— 1896. *Z. physiol. Chem.* 22:435.
279. —— 1898. *Z. physiol. Chem.* 24:18.
280. —— 1898. *Land. Jahrb.* 27:503.
281. —— 1900. *Z. physiol. Chem.* 30:241.
282. —— 1902. *Land. Vers. Sta.* 56:97.
283. —— 1902. *Land. Vers. Sta.* 56:293.
284. —— 1906. *Z. physiol. Chem.* 47:507.
285. —— 1906. *Land. Jahrb.* 35:621.
286. —— 1911. *Z. physiol. Chem.* 65:431.
287. —— and Barbieri 1877. *Ber.* 10:199; *Land. Jahrb.* 6:681.
288. —— and Barbieri 1879. *Ber.* 12:1924.
289. —— and Barbieri 1880. *Land. Vers. Sta.* 24:167.
290. —— and Barbieri 1881. *Ber.* 14:1785.
291. —— and Barbieri 1883. *J. pr. Chem.* 27:337.
292. —— and Bosshard 1883. *Land. Vers. Sta.* 29:295.
293. —— and Bosshard 1885. *Z. physiol. Chem.* 9:420.
294. —— and Castoro 1903. *Z. physiol. Chem.* 38:199.
295. —— and Castoro 1904. *Z. physiol. Chem.* 43:170.
296. —— and Kisser 1889. *Land. Vers. Sta.* 36:1.
297. —— and Steiger 1887. *Z. physiol. Chem.* 11:43.
298. —— and Steiger 1886. *Ber.* 19:1177.
299. —— and Umlauft 1876. *Land. Jahrb.* 5:819.
300. —— and Winterstein 1901. *Z. physiol. Chem.* 33:547.
301. —— and Winterstein 1910. *Z. physiol. Chem.* 65:431.
302. Schulze, T. 1932. *Planta* 16:116.
303. Schützenberger 1879. *Ann. chim. phys.* (5) 16:289.
304. Schwab 1936. *Planta* 25:579.
305. Schwabe 1932. *Protoplasma* 16:397.
306. Seliwanow 1887. *Land. Vers. Sta.* 34:414.
306a. Sen and Blackman 1933. *Ann. Bot.* 47:663.

307. SHARP 1934. *An Introduction to Cytology.* 3d Ed. New York.
308. SHOREY 1897. *J. Am. Chem. Soc.* 19 : 881.
309. SIEWERT 1870–1872. *Jahresber. f. Agriculturchem.* 2 : 6.
310. SMIRNOW 1923. *Biochem. Z.* 137 : 1.
311. —— 1928. *Planta* 6 : 687.
312. SMITH and CHIBNALL 1932. *Biochem. J.* 26 : 1345.
313. SPOEHR 1919. *Carnegie Inst. Wash.* Publ. No. 287.
314. —— and McGEE 1923. *Carnegie Inst. Wash.* Publ. No. 325.
315. STARE and BAUMANN 1936. *Proc. Roy. Soc. Lond.* B 121 : 338.
316. STIEGER 1913. *Z. physiol. Chem.* 86 : 245.
317. SULLIVAN 1858. *Ann. sci. nat.* (4 Bot.) 9 : 290.
318. —— 1926. *U.S. Pub. Health Reports* 41 : No. 22, 1030.
319. SUZUKI 1897. *Bull. Coll. Agric. Tokyo* 2 : 409.
320. —— 1900–1902. *Bull. Coll. Agric. Tokyo* 4 : 1, 25.
321. —— 1900–1902. *Bull. Coll. Agric. Tokyo* 4 : 351.
322. SZENT-GYÖRGYI 1935. *Z. physiol. Chem.* 236 : 1.
323. —— 1936. *Z. physiol. Chem.* 244 : 105.
324. TREBOUX 1904. *Ber. bot. Ges.* 22 : 570.
325. TRISTRAM 1938. Unpublished data.
326. ULLRICH 1924. *Z. f. Bot.* 16 : 513.
327. VAN SLYKE 1912. *J. Biol. Chem.* 12 : 275.
328. VAUQUELIN and ROBIQUET 1806. *Ann. chim.* 57 : 88.
329. VERKADA 1938. *Chemistry and Industry* in *J. Soc. Chem. Ind.* 57 : 704.
330. VICKERY 1925. *Year Book, Carnegie Institution of Washington,* 24 : 349.
331. —— 1925. *J. Biol. Chem.* 65 : 657.
332. —— 1927. *Plant Physiol.* 2 : 303.
333. —— 1938. *Cold Spring Harbor Symposia on Quantitative Biology* 6 : 67.
334. ——, PUCHER and CLARK 1936. *Plant. Physiol.* 11 : 413.
335. ——, PUCHER, WAKEMAN and LEAVENWORTH 1933. *Carnegie Inst. Wash.* Publ. No. 445.
336. ——, PUCHER, WAKEMAN and LEAVENWORTH 1937. *Conn. Agr. Expt. Sta.* Bull. 399.
338. ——, PUCHER, WAKEMAN and LEAVENWORTH 1938. *Conn. Agr. Expt. Sta.* Bull. 407.
339. —— and SCHMIDT 1931. *Chem. Reviews,* 2 : 169.
340. VINES 1902–1910. Various papers in *Annals Bot.* 1902–1910.

341. VIRTANEN 1938. *Annals of the Agricultural College of Sweden* 5:429.
342. —— and LAINE 1935. *Nature* 136:756.
343. —— and LAINE 1936. *Biochem. J.* 30:1509.
344. —— and LAINE 1936. *Suomen Kemistil* 9B:12.
345. —— and LAINE 1937. *Suomen Kemistil* 10B:6.
346. —— and LAINE 1937. *Suomen Kemistil* 10B:24.
348. —— and LAINE 1938. *Nature* 141:748.
349. —— and LAINE 1938. *Nature* 142:165.
350. —— and TARNANEN 1932. *Biochem. Z.* 250:193.
351. WAGNER-JAUREGG and RAUEN 1935. *Z. physiol. Chem.* 233:215.
352. WARBURG and NEGELEIN 1920. *Biochem. Z.* 110:66.
353. WASSILIEFF 1901. *Land. Vers. Sta.* 55:45.
354. —— 1904. *J. Exper. Landw.* (Russian). 34.
355. —— 1908. *Ber. bot. Ges.* 26:454.
356. WEIL-MALHERBE 1936. *Biochem. J.* 30:665.
358. —— 1937. *Biochem. J.* 31:2202.
359. —— and KREBS 1935. *Biochem. J.* 29:2077.
360. WEYL 1876. *Pflüger's Arch.* 12:635.
361. —— 1877. *Z. physiol. Chem.* 1:72.
362. WINKLER 1934. *Z. physiol. Chem.* 228:50.
363. WOOLF 1929. *Biochem. J.* 23:472.
364. YEMM 1935. *Proc. Roy. Soc. Lond.* B 117:483:504.
365. —— 1937. *Proc. Roy. Soc. Lond.* B 123:243.
366. ZACHARIAS 1883. *Bot. Zeitung* 41:209.
367. ZALESKI 1898. *Ber. bot. Ges.* 16:146.
368. —— 1897. *Ber. bot. Ges.* 15:536.
369. —— and MARX 1913. *Biochem. Z.* 48:175.
370. —— and SHATKIN 1913. *Biochem. Z.* 55:72.

INDEX

Abderhalden, 97.
Acid plants (Ruhland and Wetzel), 91.
Aconitic acid, as intermediary product of citric acid oxidation, 192.
Adler, 105.
Alpha-ketoglutaric acid, *see* Ketoglutaric acid.
Amides, confusion in early literature concerning the term, 19.
    presence of, in plants with very acid saps, 91.
Amido-N, estimation of (Sachsse-Kormann), 19.
Amino acid analysis, of leaf proteins, difficulties of, 138–140, Appendices I, II.
Amino acids, accumulation of, in detached leaves, 174 ff.
    breakdown of, Schulze's first suggestion that this might give ammonia, 48.
    decarboxylation of, in plants, 90.
    in leaves, possible nutritive value of, 166, 167, 169.
    isolation of, from white lupins, 56.
    mechanism of synthesis of, in plants, 104 ff.
    oxidation of, to ketonic acids, 90.
    synthesis of, from products supplied by symbiotic bacteria, 109–111.
    from ammonia, 105 ff.
    from nitrate, 111.
Amino nitrogen, accumulation of, in K-starved barley leaves, 258, 259, 260, 261.
    donation of, 106, 107, 110.
Ammonia, as α and ω of protein metabolism, 75, 82, 103.
    as oxidation product of amino acids, 90.
    detoxication of, in plants, 100, 193, 237.
    of proteins, Nasse's views on, 14.
    oxidation of, to amino acids, 88.
    plants (Ruhland and Wetzel), 91.
    production of, during germination, 11.
    Schulze's original suggestion that this might arise through breakdown of amino acids, 48.
    synthesis of amino acids from, 105 ff.
Andrews, 131.
Arginine, accumulation of, in seedlings of yellow lupin, 62, 63.
    as primary product of protein breakdown in seedlings, 62 ff.
    discovery of, 45.
    oxidation products of, 95, 96.
Ashton, 187.
Asparaginase, presence of, in plants, 107.
Asparagine, accumulation of, in detached leaves, 173 ff., 217, 218, 220, 236, 240.
    accumulation of, influence of temperature on, 76, 78.
    in germinating lupins, 17, 18.
    in plants enriched with ammonia, 98 ff.
    in stems, 115.
    as integral part of protein molecule (Ritthausen), 14.
    availability of, for protein synthesis (Schulze), 52.
    concentration of, in lupin-seedling saps, 18.
    discovery of, 1.
    formation of, and organic acid cycle in plants, 194.
    effect of carbohydrate and light on, 103.
    in plants, Schwab's views, 203.

requires the oxidation of amino acids, 89.
in leaves, early views on origin of, 41, 42, 48.
difficulty of isolation of, 172.
in plants, analogy with urea in animal body, 4, 9, 76, 88.
widespread occurrence of, 5, 6, 10, 27, 29, 30.
in seedlings, Schulze's original views on origin of, 37 ff.
Schulze's later views on origin of, 48 ff.
secondary production of, 50, 54 ff.
isolation of, from enzymic digest of edestin, 15, 60.
non-nitrogenous precursors of, 189, 196 ff.
origin of carbon skeleton of, 224, 234–236, 240, 242.
oxalacetic acid as precursor of, 189, 190.
possible production of, at expense of malic acid in leaves, 226–229.
possible secondary production of whole molecule from protein, 91 ff.
production and respiration, early views on, 76–78.
role of, in protein metabolism, 6–12, 26 ff.
in protein regeneration, 8 ff.
in respiration (Borodin), 27.
Schulze's views on availability of, for protein synthesis, 52.
synthesis of (Piutti), 1.
by vacuum infiltration of leaves, 197 ff.
in relation to carbohydrate and fat oxidation, 193, 194.
transformation of, to succinic acid (Mercadante), 24.
transport of, in plants, 10.
Aspartase, presence of, in plants, 190.
Aspartic acid, biological relationship to fumaric acid, 190.
discovery of (Ritthausen), 13.
donation of amino-N by, 106 ff.
in leaf proteins, estimation of, 282.
Assimilation, Müller's views on role of asparagine in, 46, 47.
Atkins, 127.

Bacteria, symbiotic, synthesis of amino acids from products supplied by, 109–111.
Baernstein, 269–271.
Bailey, 139, 269, 271, 276, 277.
Ball, 127.
Baptiste, 260.
Barbieri, 33, 35, 45.
Barley, detached leaves of, changes in organic acids of, 238, 239.
detached leaves of, respiration and chemical changes in, 212 ff.
grains of, changes in nitrogenous constituents during development, 118.
leaves of, effect of manurial treatment on protein metabolism of, 254 ff.
Baumann, 43, 190.
Beets, roots of, elaboration of glutamine in, 99.
Bennet-Clark, 195.
Bente, 37.
Bergmann, 275.
Beyer, 6, 7, 8, 16, 189.
Billimoria, 113, 118, 182.
Bishop, 117, 118.
Björkstén, 113, 196.
Blish, 273, 275.
Block, 280.
Blumenthal, 269, 270, 272.
Borodin, 26 ff., 46, 47, 49, 71, 170, 172, 173, 248, 252.
modification of Pfeffer's theory, 31, 32.
Bosshard, 172.
Boussingault, 3, 4, 5, 7, 9, 94, 100, 171.
Brand, 269.
Braunstein, 106, 107, 111.
Buds, Borodin's experiments with developing, 27 ff.
Butkewitsch, 90, 176.

Cabbage leaves, lipoids of, 284, 285.

INDEX

Cahours, 5.
Campbell, 83.
Carbohydrates, as agents of protein synthesis in plants, 112, 113.
Carotene-protein ratio in leaves, 187.
Castoro, 56, 62, 63.
Channon, 131, 284, 285.
Chautard, 2, 9.
Chibnall, 58, 60, 107, 122, 125 ff., 131, 175, 178, 179, 186, 187, 204 ff., 253, 268, 275, 278, 284, 285.
Chloroplastic material, definition of term, 130.
   from spinach leaves, composition of, 132, 136–138.
      protein-lipoid ratio in, 138.
      separation of (Menke), 131 ff.
Chloroplasts, chemical composition of, 153, 154.
   lipoid fraction of, 284, 285
   proteins of, early views on role of, 183, 184.
   properties and amino acid composition of, 155, 156.
   structure of, 153.
Citric acid, amount present in tobacco leaves, 223.
   cycle (Krebs and Johnson), 193, 232 ff.
   in lupin seeds, 16.
   of detached leaves, changes in, 223, 224, 227–229, 232, 233, 238, 239, 241, 242.
   oxidation of, to α-ketoglutaric acid, 192.
   possible reason for storage of, in plants, 193–195.
   possible synthesis from malic acid in tobacco leaves, 226–229.
   synthesis of, from oxalacetic acid, 191.
Clapp, 83.
Clark, 99.
Clarke, 269, 270, 272.
Clausen, 51, 89, 90.
Cocksfoot, blades of, amino acid analysis of protein of, 117.
   lipoids of, 284, 285.
Cohn, 18.
Corbet, 111.

Cossa, 24, 25.
Crampton, 166.
"Crude protein" of forage crops, 168, 169.
Cystine, estimation of, in leaf proteins, 268 ff.
Cystine content of pasturage and wool production, 167, 168.
Cytological terms, definitions of, 129, 130.
Cytoplasmic material, definition of term, 130.
   from spinach leaves, chemical composition of, 132, 136.
   separation of (Menke), 131 ff.
Cytoplasmic proteins of leaves, amino acid analyses of, 141, 143, 144.
   ether method of preparation (Chibnall), 134 ff.
   properties of, 141, 142.

Damodaran, 60, 95, 106.
Deleano, 172, 211, 233, 246.
Demjanow, 89.
Denny, 254.
Dessaignes, 2, 9.
Detmer, 44.
Die Eiweisskörper, 13.
Dixon, 127.
Dumas, 5.

Eckerson, 111.
Edlbacher, 96.
Eggleton, 182.
Eichhorn, 16.
Emmerling, 46, 47, 49, 113, 114.
   views of, on role of proteins in protoplasm, 49.
Endres, 110.
Enzymes in vetch seedlings, discovery of, 11.
Ether, cytolytic action of, on leaves, 128.
Ether method for preparing proteins from leaves, 126–129, 134 ff.
   limitation of, 145–147.
v. Euler, 105, 107.

Fagan, 187.

Fats, metabolism of, in detached leaves, 232, 234.
oxidation of, to give succinic acid, 193.
Ferments, activity of, in plants (Borodin), 30.
Fischer, 180.
Foreman, 147–151, 282.
Foster, 285.
Fourcroy, 122.
Fowler, 164, 166.
Frey-Wyssling, 153.
Fumaric acid, biological relationship to other dicarboxylic acids, 190.
Furfuraldehyde, condensation of, in humin formation, 274 ff.

Garreau, 27.
"Gleis," 5–7, 65.
Glutamic acid, aerobic dehydrogenase of, in seedlings, 95, 106.
dehydrase, presence of, in plants, 105.
discovery of, 12.
donation of amino-N by, 106 ff.
in leaf proteins, estimation of, 282.
oxidation products of, 95.
reversible oxidative deamination of, 105, 106.
Glutaminase, presence of, in animal and plant tissues, 107, 108, 205.
Glutamine, as component of protein molecule, first reference to, 14.
availability of, for protein synthesis (Schulze), 52.
discovery of, 14, 36.
distribution of, in plants, 58, 59.
equivalent to asparagine, in protein metabolism (Schulze), 32.
formation of, and organic acid cycle in plants, 194.
in seedlings, secondary production of, 58 ff.
in vine leaves, 172.
isolation of, from enzymic digest of gliadin, 15, 60.
α-ketoglutaric acid as precursor of, 190.
origin of carbon skeleton of, in detached leaves, 234, 235, 241.
production in detached leaves, 219, 220, 224 ff., 237, 240.
production of, in plants enriched with ammonia, 98 ff.
synthesis of, from ammonium glutamate, 205.
from di-ammonium α-ketoglutarate, 207–210.
in blades of perennial rye-grass, 204 ff.
in detached leaves, 217–219, 224, 237, 240.
Glutamine synthesis and oxidation of fats and carbohydrates, 193.
Godlewski, 69, 90.
Gortner, 273–275, 277.
v. Gorup-Besanez, 11, 19, 21, 30, 34 ff., 49, 65.
on primary products of protein breakdown in seedlings, 34 ff.
Gouwentak, 250, 254.
Graham, 10.
Grass, proteins of, amino acid composition of, 156, 158, 160–162, 164, 165.
proteins of, and milk production, 166.
Green, 50, 59.
Greenhill, 204.
Gregory, 171, 254 ff.
Grover, 42, 107, 127, 142, 186, 187, 278.

Habermann, 14, 15.
Hansteen, 248.
Hartig, 5, 6, 27, 28, 29, 38, 65.
on the translocation of nitrogenous nutrients in plants, 5, 6.
Herrick, 97.
Heyl, 83.
Histidine, oxidation products of, 96.
Hlasiwetz, 14, 15.
and Habermann, on protein composition, 14.
Hoppe-Seyler, 11.
Hosaeus, 11.
Humin, formation of, on hydrolysis

INDEX

of leaf proteins, 138–141, 274 ff.
from leaf proteins, composition of, 276–278.
Hydroxylamine, role of, in amino acid synthesis in plants, 110, 111.
Hyponitrous acid, formation of, in nitrate reduction, 111.
Hypoxanthine, possible presence of, in plant proteins, 15.

Infiltration, vacuum, protein synthesis in leaves by means of, 113.
*Iso*citric acid, as intermediary product of oxidation of citric acid, 192.

Jaaback, 60.
Johnson, 191, 193, 232.
Jordan, 285.

Kassell, 269.
Kellner, 41, 48.
α-Ketoglutaric acid, acceptance of amino-N by, 106 ff.
as precursor of glutamine, 190.
dismutation of, 193.
oxidation of, to succinic acid, 192.
reversible reductive amination of, 105, 106.
Ketonic acids, as oxidation products of amino acids, 90.
decarboxylation of, by carboxylase, 95.
Kiesel, 136, 174.
Klein, 64.
Knoop, 105, 191, 192.
Knop, 37.
Krauch, 50.
Krebs, 90, 96, 107, 191, 193, 205, 232.
Kreusler, 13.
Kritzmann, 106, 107, 111.
Kultzscher, 91.
Kutscher, 95.

Laine, 107, 109, 110, 111.
Laskowsky, 76, 78.
Leavenworth, 123, 211, 221 ff.

Leaves, age of, and protein decomposition, 78.
amino acids in, possible nutritive value of, 166, 167, 169.
anaesthetic action of chloroform vapour on, 127.
asparagine in, difficulty of detection of, 172.
basic-N of, unknown constitution of, 173.
chloroplastic proteins of, amino acid analyses of, 155, 156.
properties of, 155, 156.
control of protein level in, 244 ff.
detached, accumulation of amino acids in, 174 ff.
accumulation of asparagine in, 173 ff.
changes in organic acids of, 223.
changes in protein content of, 216, 217, 224, 225.
loss of nitrogen from, 181, 182.
parallel breakdown of protein and chlorophyll in, 184–186.
production of asparagine in, 217, 218, 220, 226, 240.
proof that proteins may be concerned in respiration of, 223, 224.
protein breakdown in, 173 ff., 183 ff.
proteins that undergo decomposition in, 184 ff.
respiratory quotient in, 215, 216, 220, 237.
translocation of protein decomposition products to petiole in, 179, 180.
diurnal variation in protein content of, 253, 254.
labile character of proteins of, 171.
of barley, accumulation of amino-N in, 257.
effect of K starvation on, 256 ff.
protein metabolism and manurial treatment of, 254 ff.
parallelism between protein metabolism and respiration in, 259.

protein breakdown in, dependence on leaf age, 178.
   related to water content, 244.
protein-carotene ratio in, 187.
protein cycle in, 171, 259.
protein level in, dependence on oxygen potential, 245, 246.
protein metabolism in, regulation of (Paech), 246 ff.
proteins of, nutritive value of, 160–163.
   stability value of, 170, 171.
protoplasm of, greater complexes in, 149, 150.
protoplasmic proteins of, preparation and properties of, 151 ff., 155, 157.
reserve proteins of, nature of, 183, 188.
saps of, hydrogen-ion concentration of, 142.
   non-protein nitrogenous constituents of, 171–173.
synthesis of proteins in, 113, 246–248.
water content of, and protein level, 244.
whole protein of, 158–160.
vacuum infiltration of, 196 ff.
Lee, 270.
Legumin, transformation of, to asparagine (Pfeffer), 8.
Lemoigne, 111.
Lepeschkin, 130.
Leucine, oxidation of, 89.
Levy, 275.
Liebalt, 154.
Liebig, 5.
Light, role of, in protein regeneration from asparagine, 4, 9.
Lobb, 183.
Löw, 43, 51, 88.
Lucerne, leaves of, non-protein nitrogenous constituents of, 172, 173.
   proteins of, 123, 124, 128, 142, 146, 156, 161.
Lugg, 139, 140, 143, 156–158, 160–162, 165, 207, 267 ff.
Lupins, arginine content of seedlings of, 62, 63.

amino acids in seedlings of, 56.
chemical analyses of seedlings of, 6, 7, 15 ff., 92.
metabolism of seedlings of, 92–94.
Lysine, estimation of, in whole leaf protein, 281, 282.

MacDougal, 139.
McGee, 260.
McKee, H. S., 55.
McKee, M. C., 183.
Malic acid, amount of, in tobacco leaves, 223.
   as possible precursor of asparagine, 6, 24, 25, 226–229.
   as possible precursor of citric acid, 226–229.
   biological relationship to other dicarboxylic acids, 190.
   of detached leaves, changes in, 223, 224, 227–229, 232, 233, 238, 239, 241, 242.
   possible reason for storage of, in plants, 193–195.
Marston, 167.
Martius, 191, 192.
Marx, 95.
Maskell, 115, 254.
Mason, 115, 127, 130, 149, 254.
Mayer, 7.
Menke, 122, 129, 131 ff., 154.
Mercadante, 24.
Merlis, 57.
Methionine, estimation of, in leaf proteins, 268 ff.
Meyer, 183.
Michael, 178–181, 184–187.
Miller, 147, 275, 278.
Molisch, 183, 188.
Morris, 164, 166.
Mothes, 64, 175–178, 196 ff., 210, 244–248, 253, 254, 264.
Müller, 46, 47, 172.

Nair, 95, 106.
Nasse, 14, 46.
Nasturtium, yellowing of leaves of, 184–187.
Negelein, 182.
Nightingale, 111, 163, 183.

INDEX 303

Nitrate, mechanism of reduction of, in plants, 111.
  reversal of reduction of, to ammonia, 182, 183.
  synthesis of amino acids from, 111.
Nitrite, as reduction product of nitrate in plants, 111.
Noack, 129.
Nuccorini, 60.
Nucleus, proteins of, in leaves, 136, 188.

Oesterlin, 105.
Onions, bulbs of, protein metabolism in, 80.
Organic acid cycle, possible role of, in protein metabolism of plants, 190 ff.
Organic acids, amount of, in tobacco leaves, 223.
  in plant materials, micro-method of estimation of, 207.
  of detached leaves, changes in, 223, 238, 239.
  position of, in metabolism of barley leaves, 258, 259.
Osborne, 83, 119 ff., 133.
Oxalacetic acid, acceptance of amino-N by, 106 ff.
  as precursor of asparagine, 189, 190.
  biological relationship to other dicarboxylic acids, 190.
  oxime of, 110.
  presence of, in pea plants, 110.
Oxalic acid, amount of, in tobacco leaves, 223.

Paech, 9, 67 ff., 113, 116, 117, 246 ff.
Palladin, 51, 89.
Pasteur, 2, 9, 14.
Pasturage, cystine content of, 167–168.
Pea, unripe seeds of, non-protein nitrogenous constituents of, 114–116.
Pearsall, 113, 181, 182.
Pelouze, 1.
Pentosans, as adulterants of leaf protein preparations, 139, 140.

Perennial rye-grass, glutamine excretion from blades of, 204.
Pfeffer, 2, 5–12, 28, 29, 31, 32, 38, 42, 43, 46, 51, 65, 71, 122, 170, 174, 177.
  theory of protein metabolism, 6 ff.
Pfenninger, 114.
Pflüger, 44.
Phenylalanine, discovery of, 36, 45.
Phillis, 127, 130, 149.
Phosphatides, metabolism of, in detached leaves, 232, 234.
Phosphatidic acid, salts of, in chloroplasts, 155, 284, 285.
Phosphorus, organic, in leaf protoplasm (Foreman), 148, 149.
Pine seedlings, base analysis of, 64.
Piria, 2, 9, 121.
Piutti, 1.
Plants, mechanism of amino acid and protein synthesis in, 104 ff.
Pods, as storehouse of nitrogenous substances, 114.
Pollard, 167, 268.
Potato, tubers of, glucose content of, 33.
Prianischnikow, 2, 52, 57, 58, 66, 67, 73, 75 ff., 88, 98 ff., 112, 115, 177, 189, 237.
Proline, oxidation products of, 96.
Protein content of leaves, diurnal variation in, 253, 254.
Protein constitution, development in theory of, 12 ff.
Protein cycle in leaves (Gregory and Sen), 259 ff.
Protein factor for forage crops, 158–160.
Protein level, in leaves, control of, 244 ff.
  in leaves, dependence on oxygen potential of, 245, 246.
Protein metabolism, ammonia as α and ω of, 75, 82, 103.
  in barley leaves, effect of K starvation on, 256 ff.
  effect of manurial treatment on, 254 ff.
  in leaves, general schema for, 194.
  in onion bulbs, 80.
  in plants (Borodin), 31, 32.

in plants, regulation of (Paech), 67 ff., 246 ff.
in seedlings, asparagine as a secondary product of, 55 ff.
glutamine as a secondary product of, 58 ff.
modern interpretation of, 86, 111–113.
Pfeffer's views on, 6 ff.
Prianischnikow's views on, 76 ff.
Schulze's early views on, 37 ff.
Schulze's later views on, 45 ff., 71 ff.
Protein structure, Baumann's views of, 43.
Proteins, ammonia of, origin of, 14.
chloroplastic, breakdown of, in detached leaves, 184 ff.
properties and amino acid analyses of, 156, 159.
comparison of leaf and seed, 83.
cytoplasmic and chloroplastic, of spinach leaves, 133, 134.
migration of, in plants, 11.
of detached leaves, changes in, 173 ff., 183 ff., 216, 217, 224, 225.
of leaf cytoplasm, amino acid analyses of, 141, 143, 144.
preparation of, 131 ff.
properties of, 141, 142.
of leaf protoplasm, amino acid analysis of, 156 ff.
properties of, 155, 156.
representativeness of, 158–160.
of leaves, estimation of bases in, 280–282.
estimation of cystine in, 268 ff.
estimation of methionine in, 268 ff.
Foreman's method of preparation, 147, 148.
labile character of, 171.
nutritive value of, 160–163.
organic phosphorus in (Foreman), 148, 149.
stable nature of, 170, 171.
synthesis of, 113, 246–248.
sulphur distribution in, 268 ff.
of plants, mechanism of synthesis of, 104 ff.

respiratory combustion of (Boussingault), 4.
of seeds, decomposition of, on germination, 3 ff., 15 ff., 54 ff.
early amino acid analyses of, 13.
regeneration of, in seedlings, 9 ff., 21 ff., 65 ff., 83, 84, 111–113.
synthesis of, from ammonia, early views, 24, 25.
in leaves, 113, 246–248.
in ripening seeds, 113 ff.
Protoplasm, of leaves, complexes present in (Foreman), 150.
Protoplasmic material, definition of term, 130.
Prunty, 268.
Pucher, 99, 183, 207, 211, 221 ff.

Quastel, 190.

Rahn, 80.
Reaction 1, 105; 2, 106; 3, 107; 4, 108; 5, 108; 6, 109; 7, 110; 8, 190; 9, 191; 10, 191; 11, 192; 12, 192; 13, 194.
Respiration, chemical mechanisms of (Borodin), 27.
of detached barley leaves, chemical changes concerned in, 212 ff.
of detached leaves, proof that protein may be concerned in, 233, 234.
of detached tobacco leaves, chemical changes concerned in, 231 ff.
of detached vine leaves, chemical changes concerned in, 211, 212.
of leaves, close parallelism of protein metabolism with, 256 ff.
of seedlings, possibility that proteins are concerned in, 92–94.
Respiratory quotient in detached leaves, 215, 216, 220, 237.
Rhubarb leaves, difficulty of preparing proteins from, 145–146.
Richards, 254 ff.
Richardson, 80, 207.
Ricinus communis, secondary produc-

tion of glutamine in seedlings of, 58 ff.
Ritthausen, 5, 8, 12–14, 16, 119.
Robertson, 167.
Robiquet, 1.
Rouelle, 121, 122.
Roxas, 274 ff.
Ruhland, 90, 91, 193.
Runner bean, leaf proteins of, 125, 126.

Sachs, 11, 183.
Sachsse, 51, 57.
Sahai, 178, 179.
Said, 260.
Salomon, 15.
Saposchnikow, 248.
Schmidt, 46.
Schryver, 122, 125.
Schützenberger, 42.
Schulow, 98 ff.
Schultze, 170.
Schulze, E., 2, 5, 12, 15 ff., 31 ff., 40, 42, 48 ff., 62, 63, 66, 71 ff., 87 ff., 92 ff., 97, 112 ff., 119, 122, 172, 174, 175, 189, 252.
Schwab, 58, 59, 91, 199 ff.
Schwabe, 260.
Seedlings, asparagine as a secondary product of metabolism in, 50, 54 ff.
  asparagine production in, influence of temperature on, 76, 78.
  glutamine as a secondary product of metabolism in, 58 ff.
  grown in light, analyses at various stages of growth, 56, 79.
  protein regeneration in, 71 ff., 111–113.
  Schulze's early views on origin of asparagine in, 37 ff.
  Schulze's later views on origin of asparagine in, 48 ff.
Seeds, ripening, isolation of amino acids from, 114–116.
  ripening, protein synthesis in, 113 ff.
Seliwanow, 79.
Sen, 171, 256 ff.
Shatkin, 80.
Shorey, 55.

Siewert, 16.
Smirnov, 196, 250.
Somers, 238, 239, 242.
Spinach leaves, proteins of, 123, 131, 132, 136–138, 141.
  separation of protoplasmic material from (Menke), 131 ff.
Spoehr, 139, 260.
Stare, 190.
Steiger, 45.
Stems, accumulation of asparagine in, 115.
Stieger, 58.
Succinic acid, biological relationship to other dicarboxylic acids, 190.
  early views on relationship of asparagine to, 24.
  possible production of, from amino acids, 95–97.
Sugars, changes in, during respiration of detached leaves, 211–213, 215, 216, 221, 222.
Sullivan, 2, 9, 268.
Suzuki, 62, 64, 90, 98, 246.
v. Szent-Györgyi, 190.

Tarnanen, 190.
Tauböck, 64.
Templeman, 256 ff.
Tobacco, detached leaves of, chemical changes in, 221 ff.
  detached leaves of, respiration of, 231 ff.
  detached stalks of, chemical changes in, 237, 238.
Tomato plants, sulphur deficiency in, 163, 164.
Translocation of nitrogenous substances in plants, 115, 116.
Treboux, 72.
Tristram, 63, 83, 156, 159–162, 165, 280–282.
"True protein" of forage crops, 168, 169.
Tryptophan, estimation of, in leaf proteins, 273.
Tryptophan content of legumes, 166, 167.
Tyrosine, detection of, in plants (Borodin), 30.

estimation of, in leaf proteins, 273.

Ullrich, 184.

Vacuolar protein, from spinach leaves, 136, 141.
Vacuolar solutes, difficulty of determining total amount of, 149.
Valine, first isolation of, from plants, 36, 45.
Vauquelin, 1.
Verkade, 193, 234.
Vetch, seedlings of, v. Gorup-Besanez's researches with, 34 ff.
Vickery, 2, 46, 55, 64, 98, 99, 108, 169, 172, 173, 183, 193, 205, 207, 211, 221 ff., 251, 265.
Vine, detached leaves of, respiration and chemical changes in, 211, 212.
Virtanen, 107, 109–111, 190.
Vitaids (Lepeschkin), 130.
Vitalism, influence of, on current opinion of protein metabolism, 44, 45.

Wakeman, 123, 133, 211, 221 ff., 251.
Warburg, 182.
Wassilieff, 56, 114.
Weil-Malherbe, 96, 105, 219.
Westall, 58, 61, 206.
Wetzel, 90, 91, 193.
Weyl, 5.
Wheat, developing grain of, 116, 117.
Winterstein, 62, 114, 116.
Wolf, 37.
Wool production in sheep, cystine content of pasturage and, 167–168.
Woolf, 190.
Wright, 164, 166.

Xanthine, possible presence of, in proteins, 15.

Yemm, 211 ff., 238, 239, 242, 251, 264.

Zacharias, 183.
Zaleski, 80, 95, 246.

SILLIMAN MEMORIAL LECTURES
PUBLISHED BY YALE UNIVERSITY PRESS

ELECTRICITY AND MATTER. By Joseph John Thomson, D.Sc., LL.D., Ph.D., F.R.S., Fellow of Trinity College and Cavendish Professor of Experimental Physics, Cambridge University. (Out of print.)

THE INTEGRATIVE ACTION OF THE NERVOUS SYSTEM. By Charles S. Sherrington, D.Sc., M.D., Hon. LL.D. Tor., F.R.S., Holt Professor of Physiology, University of Liverpool. (Out of print.)

EXPERIMENTAL AND THEORETICAL APPLICATIONS OF THERMODYNAMICS TO CHEMISTRY. By Dr. Walter Nernst, Professor and Director of the Institute of Physical Chemistry in the University of Berlin.

RADIOACTIVE TRANSFORMATIONS. By Ernest Rutherford, D.Sc., LL.D., F.R.S., Macdonald Professor of Physics, McGill University. (Out of print.)

THEORIES OF SOLUTIONS. By Svante Arrhenius, Ph.D., Sc.D., M.D., Director of the Physico-Chemical Department of the Nobel Institute, Stockholm, Sweden. (Fourth printing.)

IRRITABILITY. A Physiological Analysis of the General Effect of Stimuli in Living Substances. By Max Verworn, M.D., Ph.D., Professor at Bonn Physiological Institute. (Second printing.)

STELLAR MOTIONS. With Special Reference to Motions Determined by Means of the Spectrograph. By William Wallace Campbell, Sc.D., LL.D., Director of the Lick Observatory, University of California. (Second printing.)

PROBLEMS OF GENETICS. By William Bateson, M.A., F.R.S., Director of the John Innes Horticultural Institution, Merton Park, Surrey, England. (Out of print.)

THE PROBLEM OF VOLCANISM. By Joseph Paxson Iddings, Ph.B., Sc.D. (Second printing.)

PROBLEMS OF AMERICAN GEOLOGY. By William North Rice, Frank D. Adams, Arthur P. Coleman, Charles D. Walcott, Waldemar Lindgren, Frederick Leslie Ransome and William D. Matthew. (Second printing.)

ORGANISM AND ENVIRONMENT AS ILLUSTRATED BY THE PHYSIOLOGY OF BREATHING. By John Scott Haldane, M.A., M.D., F.R.S., Hon. LL.D. Birm. and Edin., Fellow of New College, Oxford; Honorary Professor, Birmingham University. (Third printing.)

A CENTURY OF SCIENCE IN AMERICA. With Special Reference to the American Journal of Science, 1818-1918. By Edward Salisbury Dana, Charles Schuchert, Herbert E. Gregory, Joseph Barrell, George Otis Smith, Richard Swann Lull, Louis V. Pirsson, William E. Ford, R. B. Sosman, Horace L. Wells, Harry W. Foote, Leigh Page, Wesley R. Coe and George L. Goodale. (Out of print.)

A TREATISE ON THE TRANSFORMATION OF THE INTESTINAL FLORA WITH SPECIAL REFERENCE TO THE IMPLANTATION OF BACILLUS ACIDOPHILUS. By Leo F. Rettger, Professor of Bacteriology, Yale University, and Harry A. Cheplin, Seessel Fellow in Bacteriology, Yale University. (Out of print.)

THE EVOLUTION OF MODERN MEDICINE. By Sir William Osler, Bart., M.D., F.R.S. (Fifth printing.)

RESPIRATION. By J. S. Haldane, M.A., M.D., F.R.S., Hon. LL.D. Birm. and Edin., Fellow of New College, Oxford; Honorary Professor, Birmingham University, and J. G. Priestley. (New edition.)

AFTER LIFE IN ROMAN PAGANISM. By Franz Cumont. (Second printing.)

THE ANATOMY AND PHYSIOLOGY OF CAPILLARIES. By August Krogh, Ph.D., LL.D., Professor of Zoö-physiology, Copenhagen University. (Enlarged and revised edition, third printing.)

LECTURES ON CAUCHY'S PROBLEM IN LINEAR PARTIAL DIFFERENTIAL EQUATIONS. By Jacques Hadamard, LL.D., Member of the French Academy of Sciences; Foreign Honorary Member of the American Academy of Arts and Sciences.

THE THEORY OF THE GENE. By Thomas Hunt Morgan, LL.D., Sc.D., Ph.D., Professor of Biology, California Institute of Technology. (Enlarged and revised edition.)

THE ANATOMY OF SCIENCE. By Gilbert N. Lewis, Ph.D., Sc.D., Professor of Chemistry and Dean of the College of Chemistry, University of California. (Second printing.)

BLOOD. A Study in General Physiology. By Lawrence J. Henderson, A.B., M.D., Professor of Biological Chemistry in Harvard University.

ON THE MECHANISM OF OXIDATION. By Heinrich Wieland, Professor of Organic Chemistry, University of Munich. (Second printing.)

MOLECULAR HYDROGEN AND ITS SPECTRUM. By Owen Willans Richardson, Yarrow Research Professor of the Royal Society, King's College, London.

THE CHANGING WORLD OF THE ICE AGE. By Reginald Aldworth Daly, Sturgis Hooper Professor of Geology, Harvard University. (Second printing.)

THE REALM OF THE NEBULAE. By Edwin Hubble. (Second printing.)

EMBRYONIC DEVELOPMENT AND INDUCTION. By Hans Spemann, Professor Emeritus of Zoölogy, University of Freiburg im Breisgau.

UNIVERSITY OF CALIFORNIA
MEDICAL SCHOOL LIBRARY

THIS BOOK IS DUE ON THE LAST DATE STAMPED BELOW

Books not returned on time are subject to a fine of 50c per volume after the third day overdue, increasing to $1.00 per volume after the sixth day. Books not in demand may be renewed if application is made before expiration of loan period.

AUG 26 1952
Sep. 76 "
JUL 7 1955
FEB 17 1958

3m-8,'38(3929s)